ERWIN BAUER'S
HORNED
AND ANTLERED
GAME

HORNED

ERWIN BAUER'S
AND ANTLERED GAME

By Erwin A. Bauer

Photos by ERWIN A. and PEGGY BAUER

PUBLISHED BY OUTDOOR LIFE BOOKS, NEW YORK

MOOSE: TITLE PAGES
The first light of morning, during the
October rut in Wyoming, finds a bull moose
on the prowl.

RAM: THREE PAGES PRIOR
Wild sheep, such as this Rocky Mountain
bighorn ram, are reigning mountain
monarchs—the rest of the world far below.

Copyright © 1986 by Erwin A. Bauer

Published by
  Outdoor Life Books
  Times Mirror Magazines, Inc.
  380 Madison Avenue
  New York, NY 10017

Distributed to the trade by
  Stackpole Books
  Cameron and Kelker Streets
  P.O. Box 1831
  Harrisburg, PA 17105

Produced by Soderstrom Publishing Group Inc.
Book design by Nai Chang

Typography by Westchester Book Composition Inc.

Library of Congress Cataloging-in-Publication Data
Bauer, Erwin A.
  Erwin Bauer's horned and antlered game

  Includes index.
  1. Cervidae. 1. Bovidae. 3. Deer—North America.
4. Mountain sheep. 5. Mammals—North America
I. Bauer, Peggy. II. Title. III. Title: Horned and antlered game
QL737.U55B384 1986   599.73′57   86–61302
ISBN 0–943822–73–4

Manufactured in the United States of America

CONTENTS

PREFACE

One cold morning, while I was writing the moose chapter in this book, I heard a crunching in the deep snow just outside my home office window. I went to the window and spotted the cruncher—a cow moose just 6 feet away—vigorously rubbing her flank against a post of our deck.

When she finished rubbing, she walked a little farther, pawed an oval in the snow and bedded down there. After I'd finished writing the chapter, she was still resting there. Although moose are fairly common in our yard (we have had as many as five at one time), this visit, while I was writing on moose, seemed quite a coincidence.

As wildlife writers and photographers, Peggy and I are lucky to live in northwestern Wyoming. Like moose, mule deer also show up occasionally. They are more nervous than moose around our house and are intimidated by neighborhood dogs. In fact, moose ignore dogs. Mostly during late autumn nights, elk file through and past our place en route from summer ranges in the Tetons to winter pasture on the National Elk Refuge. From May to September, we can usually find a herd of bison and scattered bands of pronghorn just a short drive away. And in summer, a day's hike will take us into bighorn sheep country. In winter, bighorns are a lot easier to find because they move to lower elevations.

Although our situation is unusually good, most North Americans also live within range of big-game animals. Nowhere else on earth is so much big game within such easy access as in the United States and Canada. Even city dwellers can often spot whitetail deer just beyond city limits. For example, we have photographed whitetails in Everglades National Park just outside Miami, via the Florida Turnpike. Last winter we studied desert bighorn rams, an uncommon and elusive species, less than a half day's drive from Los Angeles and San Diego. So that part of our great wildlife heritage still somehow survives.

But I would be foolish to be optimistic about the future of North American wildlife. Especially the larger animals need wilderness to prosper. Such species as sheep, elk, and caribou need large tracts of wilderness. But lately the only growth industries are those that chew up beautiful wilderness, leaving only ugliness in the wake. As I write this, we are enduring a period in which there is little trace of any conservation ethic in the White House. Today, little money can be spared for environmental research and restoration, for wildlife management, and for national parks. Pollution and poisonous wastes are overlooked. If acid rain totally wipes out the northeastern forests, and the deer disappear, so what? We still have the vast West, haven't we?

Yes and no. All those public lands of the West are being logged too much and slashed by roads nobody needs. Maybe the worst damage of all to wildlife comes from the overgrazing of public lands by livestock

that is permitted everywhere. Both the timber rights and the grazing rights are now so cheap for a privileged few that they constitute a national disgrace. Americans are actually paying taxes to support the steady depletion of natural resources.

There may be hope though because more Americans are becoming increasingly distressed with the rape of wilderness and other public lands. During recent years, memberships have increased in organizations such as the National Wildlife Federation, the National Audubon Society, the Sierra Club, and the Wilderness Society. All of these organizations are committed to saving our last wild lands and the wild creatures that live on them. I urge you to join at least one of these groups.

Erwin A. Bauer
Teton Village, Wyoming

This large Shiras bull is one of many moose Peggy and I routinely encounter each year in our yard.

1.
ELK

Every autumn toward the tag end of September, Peggy and I try to spend a week or so in Yellowstone National Park. By then, summer's myriad tourists have retreated homeward, the weather is cool and crisp, and we have the place to ourselves once more. Yellowstone is never more gorgeous than when the quaking aspens turn yellow, and when the grassy meadows become golden again.

But we go to Yellowstone mostly because the end of September is the rutting season for the park's 20,000 or so elk. Then no other sanctuary contains so many American elk. Nowhere else can the shrill, calliope bugling of testy bulls be so widely heard, even along the main park roads and down from the elk's more remote summer pastures.

We make our headquarters at either Mammoth, Madison, or Norris public campground, within a short drive or hike to most of the rutting activity. That is important, because during the peak of the rut, activity begins before dawn.

It was well before daylight one recent morning when we could hear splashing in the Gibbon River and a bull bugling a challenge just beyond where we were awakening in our camper van. Sitting up in my sleeping bag, my breath heavy and white on the cold air, I pulled on an extra pair of heavy woolen socks and slipped into damp hipboots. Now two bulls were bugling back and forth, while a pot of coffee brewed on a small propane camp stove and the first pale light of morning revealed a frosty landscape. After wolfing several doughnuts along with the coffee, Peggy and I picked up cameras and tripods and headed toward the musical elk. A weak sun was blinking through the lodgepole pines on the eastern side of Gibbon meadow.

I was so intent on locating the elk that I did not at first notice the trail another animal had made a few minutes before through the brittle and frosted knee-high grass. Judging by the width of it, the passerby could only have been a grizzly bear. Suddenly that added a new dimension; now we had to watch for the bear as well as the elk.

The elk were standing about 100 yards beyond the Gibbon River. Some were grazing. A splendid dark bull, the ivory tips of his antlers clearly visible, stood in the center of the herd. Tilting the heavy antlers far back while staring toward the sky, the bull whistled once more just

On a crisp morning in early fall, this cow nurses her now sizable calf. Backlit by the sun, scenes such as this can make rising before dawn especially worthwhile.

PAGE 10
During the rutting season in elk country, the meadows and forest openings echo and re-echo with the strange, calliope bugling of the bulls.

Between bouts with would-be suitors, a harem bull such as this one leads an exasperating life, continually needing to round up cows that stray.

as we were wading across the cold, clear river current. I have heard that call hundreds of times before, but it never fails to raise the hairs on the back of my neck.

We maneuvered about slowly, cautiously. We were trying to get a better camera angle on the bull and also take advantage of the improving morning sunlight. What we saw through telephoto lenses was a classic scene of one of the world's most impressive creatures. These elk are long used to seeing Yellowstone tourists, and they pay little attention to photographers who keep a respectable distance, as we did.

One of the cows followed by her calf of the previous spring tried to drift away from the harem, but the bull quickly bounded in pursuit and threatened the female back into the group. Another cow also strayed and the bull rounded up that one, too. Soon a few cows began to bed down as they could feel the first warmth of the September sun. The next thing we knew, our bull was galloping away to meet a second bull that had appeared from the west.

The harem bull lunged at the new arrival just as he emerged, dripping, from crossing the Gibbon River. Although we could not clearly see the crashing of antlers and the hoofs digging the turf, we could certainly hear them. Almost before we could aim cameras in that direction, the issue was already settled and the interloping bull was running away. The winner bugled once more, returned to his harem, and seemed to count the cows before bedding down among them. Only then did we remember the grizzly bear that was also in the vicinity. This was a typical late September morning for Yellowstone's elk.

Origin of the American Elk

The American elk or wapiti, *Cervus canadensis*, is another of those ungulates that migrated here from Asia. It is believed to have originated in Tibet or the Tien Shan Mountains of western China and dispersed in opposite directions from there. The animals moving westward became the red deer of Europe, the Soviet Union, and regions as far south as Iran. Some walked south to India and became today's barasingas. Those that trekked eastward crossed the now-nonexistent Bering land bridge were probably already widespread by the Stone Age.

Except for moose, elk are the largest members of the North American deer family, and they were probably the most widespread of all hoofed animals when Columbus reached the New World. They existed from central California to Pennsylvania and the Carolinas, and from Mexico to all across southern Canada. Wapiti shared the prairies with bison and pronghorns. Apparently they were absent only in the Great Basin (Nevada, and portions of Utah, Arizona, eastern Oregon), the deep South and Gulf Coast, and some of New England. The explorer Jacques Cartier saw elk along the St. Lawrence River in 1535, and they were abundant in the northern Great Lakes region. But the elk herds east of the Mississippi had vanished in the face of expanding human settlements of the early 1800s. Lewis and Clark did not even mention the animals until their expedition reached western Missouri.

Adaptability

The early distribution reveals that elk could live in a variety of habitats, and this ability is evident today. After being eliminated from Michigan long before, seven elk were transported by boxcar from Yellowstone and reintroduced to the northern part of the Lower Peninsula in 1918. The herd reached 1,500 animals by 1960, but had leveled at about 1,000 in 1984 when a hunting season was held. Since an adult elk eats as much every day as three full-grown whitetails, the Michigan herd was beginning to cause excessive damage to forests and farms.

Elk also live along with bison east of the Rockies in the Wichita Mountains, Oklahoma, the Black Hills of South Dakota, and on Afognak and Raspberry islands, Alaska, where they never roamed naturally. Eight animals—three males and five females—from Washington's Olympic Peninsula were released on the Alaskan islands in 1928, and the resultant herd is now estimated at 1,500.

Numbers and Range Today

Today elk survive principally in two regions of North America: in the coastal mountains of the Pacific Northwest, and—in greatest numbers —in most ranges of the Rocky Mountains from New Mexico northward to Alberta and British Columbia. The range of elk coincides roughly with the range of mule deer and blacktails. Wapiti need wilderness, large

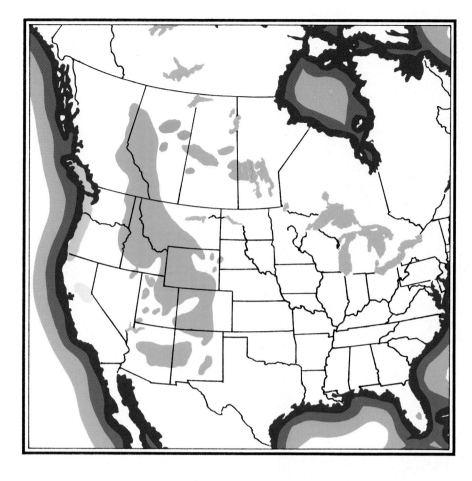

tracts of it, and today the Rocky Mountains contain the largest bloc of
wild lands left in the lower 48 states.

Originally there were six subspecies of elk, of which five survive.
Merriam's elk, once found from North Texas to Arizona, is gone. The
smallest, the Tule elk of west central California, exists in a couple of
remnant herds in the Kern Valley. The Rocky Mountain elk is most numerous and occupies the widest range. The largest of all, the Roosevelt
or Olympic elk, wanders the damp, lush coastal areas of the Pacific
Northwest.

Physique

Many hunters around the world regard our wapiti among the hand-
somest, most desirable of trophies. With antlers that may well reach 60
pounds, with multiple points that spread as wide as 48 inches from tip to
tip, majestic is a suitable adjective.

A bull wapiti is generally figured to be mature when he is 4½ or
perhaps 5½ years old, a "royal" elk for trophy purposes if his antlers
bear six points per side. Bulls with larger racks than that, with seven
points, were probably more common than they are today, but were unusual animals nonetheless. One of the largest elk I have ever seen and
photographed was only a five pointer. But the main beams were abnor-
mally heavy and long enough to brush against the animal's rump when it

tilted its head to bugle. Peggy and I have also filmed one Olympic elk with 15 total points, seven on one side and eight on the other. This was a truly awesome animal, still in the velvet, but with rack fully developed in late August when we saw it.

Because of their economic importance as big-game trophy animals, wapiti have benefited from a good bit of tax-funded scientific study, and from special efforts to protect the Rocky Mountain herds. Continuing investigation and monitoring of the herds at the National Elk Refuge, Jackson Hole, Wyoming, have revealed some hard facts. Bull elk stand 4½ to 5 feet tall at the shoulder, and weigh between 550 to 800 pounds depending on age and season. Cows weigh between 450 and 600 pounds. An elk's color varies from light brown—with darker head and legs, and lighter rump patch—from summer through fall, to grayish in winter.

I have read reports of bulls of the Olympic subspecies on Afognak Island weighing 1,000 and 1,100 pounds, field-dressed, on logging company scales. These same animals would have weighed about 1,400 to 1,500 pounds on the hoof.

Bull elk can grow massive antlers in just a short summer season. This Roosevelt bull, still in the velvet and photographed in Washington, has 15 total points. A 12-pointer —six per side—is considered a "royal."

The Jackson Hole Herd

The history of the Jackson Hole elk herd is worth relating because it is both typical and, in recent years, controversial. Archeologists believe that wapiti have wintered in this high, wide valley of the Snake River, northwestern Wyoming, for at least five centuries. As many as 25,000 animals may have wintered in the Hole when white men first began to settle the region in about 1900. Thousands of these elk may also have starved, especially during severe winters. The elk gathered here from the Tetons, from as far away as summer ranges in Yellowstone Park and from what today is the Teton Wilderness of the Bridger-Teton National Forest.

Biologists do not know whether today's Jackson Hole elk migrations from summer to winter range were as long as or longer than they were in the past. But this much is fairly certain: any migration generally southward was effectively cut off by ranching operations and by the founding of the city of Jackson. The National Elk Refuge, established in 1912, saved the herd by feeding the animals over the winter and containing them. About 8,000 elk currently winter on the Refuge, which covers 24,000 acres. The winter herd consists of 20 percent bulls, 20 percent calves, and 60 percent cows. They begin to arrive in November and stay about six months.

While there is ample summer range to support a large elk population in the Rocky Mountains, elk depend on man to get them through even an average winter because traditional winter ranges have been usurped

By September in Yellowstone, elk have descended from higher summer range for the onset of the rut.

ABOVE
During winter in Yellowstone
Park, wintering elk can be
found near geysers, hot
springs, and rivers that do
not freeze. There, the snow
pack is not too deep and
some forage is usually available at water's edge or just
beneath the snow.

LEFT
Each year, anywhere from
5,000 to 8,000 elk spend the
winter at the National Elk
Refuge in Jackson Hole,
Wyoming. Part of the herd
waits here for the daily ration
of hay or alfalfa pellets that
will be delivered. Only a
fraction of these elk could
survive the winter without the
existence of the Refuge.

by towns and shopping centers. At the National Elk Refuge, supplemental feeding of pellets or cubes of pressed alfalfa takes place during two months of the most severe weather. The animals are fed about seven pounds per animal or 30 tons per day total of food. It may not be necessary to feed the elk that much or even at all during Wyoming's rare mild winters, because they can subsist on the vast managed pastures ungrazed by domestic livestock. In general, however, winter is the critical period in an elk herd's survival.

The National Elk Refuge is more than a vital winter sanctuary. It is also anyone's best bet to see at very close distance hundreds and often thousands of elk at one time. Daily from about Christmas until early April, a concessionaire carries Refuge visitors on horsedrawn sleighs right among the massed animals. The sight of so many antlered bulls at once, with the Teton Range looming in the background, is among America's grandest wildlife spectacles. It is one that must have captivated the late renowned naturalist, writer, and wildlife artist Olaus Murie, who studied the Jackson Hole elk for 15 years, while collecting data on elk from other regions.

Born in 1899, Murie is still recognized as the foremost authority on our continental elk. He began his scientific career at age 15 on an expedition to Hudson Bay to collect specimens for the Carnegie Museum. By the time he came to Moose, Wyoming, in 1946 with his wife Margaret, Murie had an extraordinary amount of experience studying and sketching various animals throughout the world. He went on to compile most of what is known about elk.

Murie also found time to become a founder of the Wilderness Soci-

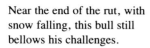

Near the end of the rut, with snow falling, this bull still bellows his challenges.

Even during midwinter, long after the rutting season, I have found that bull elk will occasionally pause in their feeding to cross antlers. The struggle for ranking, even for choice bits of feed, goes on the year-round.

ety, which today is a strong voice for saving the last precious wild places left in the United States. Most of the large mammals considered in this book would not be able to survive very long without such places.

Behaviors

Like the caribou, the elk is a herd animal, and many of its responses are based on this fact, or instinct. The extent of the herding, however, varies from season to season. Mature bulls remain aloof from cows and calves most of the year. During the calving season the cows are scattered widely in small groups all across summer range. But once the spotted calves are able to travel, the females reassemble into larger and larger groups. I have found herds of 25 to 30 animals to be commonplace in Wyoming and Montana in July. One August in Yellowstone Park, Peggy and I tallied over 300 in a herd northeast of Mt. Washburn.

Elk are shy and suspicious of people unless accustomed to them from birth, as they are in several national parks. During the summer a herd of elk can be approached nearer on horseback than on foot. While exploring the matchless wild Yellowstone backcountry from the headwaters of the Thorofare River to the Trident Plateau with pack trip outfitters Ralph and Candace Miller, we had ample evidence that even large herds could be approached within photographic range by horsemen. The Millers are based in Cooke City, Montana, and conduct trail rides—saddle safaris, actually—into the remotest areas of Yellowstone throughout every summer. Invariably these are superb adventures, during which a rider sees moose and mule deer, grizzlies and smaller wildlife, as well as the abundant elk.

STAMPEDES. The herding instinct can cause elk to panic and suddenly stampede at some sign of danger. The danger may be perceived by only one or two members of a herd, but all will break and run anyway. A person suddenly appearing on foot can cause a stampede and so can a grizzly bear, although elk seem less alarmed of bruins. The worst stampede Peggy and I ever saw was caused by free-roaming dogs in winter on the National Elk Refuge. In a matter of seconds, a herd of a thousand animals was triggered into running by a house dog that had wandered onto the refuge from adjacent Jackson. Rangers have had to shoot such dogs, which persistently harass winter-weakened elk.

By contrast, in Yellowstone's Hayden Valley in September 1969, I watched from the park road as a grizzly sow and cub walked through the center of a herd of 48 cows and calves. The animals did not run, but merely divided enough to give the bears a generous corridor to walk through. Nor did the bears even seem to notice the elk, which were soon back to normal grazing.

I have noticed another trait of elk that might be called, simply, stubbornness. This is a trait shared with pronghorn antelope. Once they decide on a course of exit or escape, they will not turn away no matter what the obstacle or situation. A number of times I have been riding past an elk herd when I was suddenly discovered. The elk would break toward the nearest high ground and cover, even if it meant crossing directly in my path, and even if I spurred my horse to try to intercept them. The running animals would never change their direction to avoid a close brush.

Diet

Except to say that elk will eat many if not most of the plants wherever they live, either grazing or browsing, the subject of elk food is far too vast to consider in detail here. The elk of one mountain range will subsist on plants that may not even grow on an adjacent or nearby range. Those interested in a detailed study of elk diet should read Olaus Murie's *The Elk of North America*. In it Murie lists, plant species by plant species, all the food preferences of elk, which he determined by direct observation of feeding animals, as well as by examining grazing areas and even stomach contents.

Why Elk Bugle

It is easy to understand why so many wildlife artists paint bull elk with extravagant antlers, standing amid females in an alpine meadow, bugling. Few sights and sounds in all the outdoors are more stirring. But while a bull's antlers have definite use during the rut, why an elk bugles remains a mystery, a subject of myths and misconceptions.

One myth, frequently heard, is that the size and age of a bull can be determined by the sound of his voice, its volume, or both. Not true. While yearling bulls can be separated from old timers by their bugling, a mature four- or five-pointer can scream as loud, or louder, and with as much penetration as that one with the largest rack in the Rockies. On the other hand, he may not. Volume simply is not achieved in direct proportion to body, antler size, or age.

During the rut an elk bull's behavior is lively and aggressive. A male will alternately tear up the turf with his antlers, bugle, and then scrape the bark from a tree in his path. Peggy and I often locate the bulls by first finding the skinned-up trees.

According to Canadian animal behaviorist Valerius Geist, bugling is a form of male advertising, and harem bulls—which on average are larger and more powerful than bulls without harems—try to "out-advertise" or "one-up" their competitors during the rut. But what happens, by the same reasoning, is that harem bulls tend to bugle more often, perhaps every minute or two during peak mornings in the rut. These frequent buglers may not have sounded any larger than their rivals, but odds are good that they were instantly recognized by other elk.

The Rut

Speaking of myths, many outdoorsmen believe that unseasonably hot weather delays the breeding season, while an early cold spell may hasten its arrival. But the truth is that hot or cold weather has little bearing on the rut. It is determined instead by photoperiodism. Every September the amount of daylight on about the same date triggers hormonal changes in the species. Photoperiodism also determines what time of the year the other animals in this book will breed.

It is true that during an exceedingly hot and dry autumn, the rut may not seem as dramatic and noisy as during cooler September weather. But this is because the animals spend daytimes in the shade of forests and are most active at night. More than once when camped in the high country, I have heard bulls bugling all night long. In the northern Rockies wapiti bulls seem most high strung during the early rut, the

The three photos on this page show a splendid bull in Alberta, beside himself with passion. He has isolated and proceeded to entertain a cow about to come into estrus. His was a virtuoso performance to watch.

PAGES 26 & 27
For weeks during the rut, in all weather, a harem bull has a full-time job keeping his group intact. He constantly rounds up wandering cows and drives away rival bulls. By season's end, most bulls are thin and exhausted and require a few weeks of moderate weather to recuperate before taking on winter.

At the end of the breeding season all becomes fairly peaceful again. Bulls that have been serious rival contenders during past weeks now bed down together in a bright frosty meadow to indulge in a morning of indolence.

peak of which falls between the middle and end of September, whether or not you hear much bugling. Most of the females have been bred by mid-October, by which time the harems have disintegrated. But we have heard an occasional bull bugling as late as November in Yellowstone Park, as well as in Texas where elk have been restocked on a few ranches.

Also I have read elk hunting accounts that indicate the species is most active during one phase of the moon or another. The most popular theory—or myth, in this case—is that rutting is more active or more lively in daylight hours during the dark of the moon. But biologists who have followed radio-collared elk both day and night and at all phases of the moon during the rut report that activity is just about the same. The elk see very well in darkness and are not inconvenienced by it.

Births and the Early Months

Elk cows are not nearly as prolific as smaller North American deer. Only once in two decades of observation have Peggy and I seen a cow with twins. In one study of Wyoming elk, only two females in 600 bore twins. But this low production—compared to whitetails and mule deer, which often have two fawns—may be compensated for by a cow's greater ability to defend calves from predators. A larger percentage of elk calves than deer fawns reach their first birthday simply because their mothers are larger and more formidable.

THE NEWBORN. A newborn calf is long-legged, ungainly, and more richly colored than its mother. It is tawny, yellow-spotted with a yellow-brown rump. I've noticed that the base color of wapiti calves varies a great deal from light reddish to very dark, from area to area. The calves are extremely difficult to see when lying motionless in good cover. In the northern Rockies a newborn calf will already weigh 30 or 40 pounds, and the calving actually takes place at 7,000 feet elevation or higher, which is about the upper limit of the elk winter range.

Although wobbly and uncertain, new calves are able to stand well enough to nurse soon after birth. But mobility is limited for several days to staggering just a few yards. The first week of life is largely spent lying down, hiding, not far from where the mother is feeding. After that the average calf can walk fairly well, gaining strength daily until it soon can run faster than a man. Not much later it will be able to outrun a grizzly.

THE MOTHER. There is a stronger bond between some cows and calves than between others. In Yellowstone Park we have watched some young elk beside their mothers closely at all times and wherever the mothers traveled. Some cows may even show interest in a shrilly bleating calf not its own. But quite often we have watched females kick and try to drive away lonely calves that did not belong to them.

CALVES AT PLAY. Play is a common activity of calves, and it is pleasant to watch the very young ones gamboling awkwardly over a

This calf is several days old and is just now able to follow its mother fairly well. But the calf soon attempted to follow its mother across the Gardiner River, which required almost more strength than it possessed. The calf was nearly swept away downstream before managing to reach safety.

bright green meadow of early summer. Watch closely and you may also see certain stronger calves already trying to exert dominance over other calves at the tender age of one month. But once during a pack trip with Ralph Miller, in a morning alpine meadow bathed in sunshine, we watched a whole band of cows and calves playing around a small pond that until a few days before had been a lingering patch of snow. The cows as well as their offspring would run around the water and through it, prancing and splashing, as if the activity gave them great enjoyment. Earlier I had read a description of a very similar incident in Olaus Murie's *The Elk of North America.*

Fatalities

Although a robust, tenacious species, elk are subject to their share of insect pests, parasites, ailments, and diseases. Like some humans, some elk may also be accident prone. Blindness is not a common affliction of elk, yet it occurs with surprising regularity. Most often an animal is blinded in only one eye, probably caused by a pointed branch or limb when the elk is running through timber. But I have seen bulls with eyes injured or missing as the probable result of dueling during the rut.

For several autumns during the 1960s a one-eyed, fourteen-point male was the dominant herd bull on Madison Meadows in Yellowstone Park. Despite the vision handicap, the bull had no trouble herding its cows and driving other bulls away. Even more remarkable is that a number of blind cows—apparently blind for a long time—have been

Stream and river crossings can take a toll in small calves. These are doing okay. But younger calves or heavier rapids can spell trouble.

This bull managed to snap off braided wire fencing without hurting himself badly. He eventually shed the wire along with his antlers. Other bulls are not always so lucky, lacerating themselves severely in barbed wire or otherwise becoming hopelessly entangled.

identified among the healthy herds on the National Elk Refuge. Somehow these animals manage to find their way from summer to winter range without being able to see.

Drowning causes some elk mortalities, although elk are good and willing swimmers. One spring, Peggy and I watched two cows easily wade across the Gardiner River in Yellowstone Park, even though the cold, snowmelt water was about belly deep. Their two calves following them hesitated, but nonetheless also tried to make the crossing. Both eventually made it, although both were swept far downstream before they could again feel firm gravel underfoot. One barely escaped being washed underneath a deadfall where it would have perished. I suspect that many calves are lost when crossing such swollen streams.

Most drownings occur during spring and fall migrations, when elk must break through a crust of ice before plunging into a current, after which they have trouble breaking out again. In the fall of 1930, according to Olaus Murie, a band of elk attempted to cross the Snake River in the northern part of Jackson Hole. Sixty head drowned in the same place that other elk had died before, and since.

Barbed wire fences are a serious danger to elk. Elk are good, high jumpers and have no trouble clearing fences or barriers to reach rancher's haystacks during the depths of winter. But they still have a habit of becoming hopelessly entangled in low wire fences. We have in our files two photos of bulls and barbed wire. One hooked its antlers into the fencing and, after a furious struggle, must have destroyed a 75-yard section of it. But the animal probably died a lingering death in the long interval before it was found by a cowpoke searching for stray dogies.

Although elk readily cross rivers in winter, the water drains precious body heat and energy reserves.

The other bull we filmed had somehow broken free, but was carrying a mass of twisted barbed wire in its rack.

The world of the American wapiti is breathtakingly beautiful at any time. But even what passes for the gentleness and peace of summer is deceptive. And winters can be just plain savage for elk, which must tolerate winter, rather than fly away or hibernate. In March 1983, for example, a thunderous avalanche triggered by spring thaws in the Pecos Wilderness, New Mexico, swept 50 elk to their death in an instant.

Elk Diseases

Arthritis is common among elk and is especially noticeable among wintering animals. The knee and hip joints visibly swell and the elk are not nimble. But not all swollen joints are arthritic. The injuries to a bull's legs may also have come from rutting combat, or even from a wound by a hunter. Or the "arthritis" may be necrotic stomatitis, which Olaus Murie regarded as the most serious by far of elk diseases. It has caused heavy losses to wintering herds, especially in late winter in the congested conditions around feeding grounds established by people. Such feeding grounds, or lots, are common in the northern Rockies to sustain elk herds through winters. The National Elk Refuge is the largest of these. Necrotic stomatitis also appears in natural areas overpopulated by elk, mule deer, or both.

SYMPTOMS OF NECROTIC STOMATITIS. The first signs of necrotic stomatitis are emaciation, listlessness, drooling, and drooping of the ears. Breathing is labored. An animal may try to continue to eat, but eventually it beds down and is unable to arise. The animal is then usually dead within a day. As often as not, coyotes, eagles, ravens, and magpies will quickly consume the carcass, and the coyotes may even be blamed for killing the elk. But during the decades that I have talked with many employees of the National Elk Refuge, where coyotes daily wander among the elk herds, not one observer has ever seen coyotes attack or feed on an elk while life remained. Some of the employees have seen coyotes sitting or sleeping nearby sick or badly injured elk, as if waiting for them to die. The cause of necrotic stomatitis isn't entirely clear, nor is any surefire cure known. But, some animals do recover.

Fighting

Fighting is hazardous to a bull's health. One fall Frank Roth shot a bull south of Butte, Montana, that seemed to have a third antler growing out of its forehead. But on close examination, Frank discovered that the horn belonged to another bull. During a rutting battle, the horn had been driven completely through the skull and one inch into the roof of the mouth before being broken off. When shot, the elk was in fair condition and might even have survived the upcoming winter.

In fall 1983, Oregon State Police Trooper Russell Ellsworth was investigating a tip from a deer hunter that an elk had been poached. What he found was the site of a vicious fight in which both the attacker and the attackee elk had fought to a draw. Exhausted and in shock, the attacker succumbed and was dead when Ellsworth arrived.

That elk was a great Roosevelt bull with seven points on one antler and eight on the other, altogether a massive specimen of the species. But its adversary was even more massive; the adversary was a steel snow-gauge tower, painted red, near the headwaters of the Collawash River in the Clackamas drainage. We will never know why the bull decided to attack a steel tower, which was fabricated of 3-inch posts imbedded in concrete, and fight it to the death. The tower was built that tall and painted red so that it could be spotted and read from the air. One theory is that the bull may have mistaken the wind whistling through the structure for the challenge of another bull.

The ground around the tower was plowed up by elk hoofs, and clumps of hair were strewn about. Trooper Ellsworth found red paint and thread marks, from bolts that fastened cross arms to the tower, on the dead elk's antlers.

OTHER INCIDENTS. The preceding was only one of several bizarre incidents involving Roosevelt bull elk that same fall. Another state trooper, Gary Holm, reported the death of a second bull on the Jewell Meadows Wildlife Area in Clatsop County. This one had taken on another bull rather than a steel tower, and one whole antler had been torn, along with a part of the skull, completely from the elk's head. The power of any thrust necessary to accomplish such damage is difficult to comprehend. Somehow the winner was able to walk away from the violent scene. Trooper Holm, who lived near the area, further reported that

This young bull has lost one antler in a contest with a rival and proceeded to take out his frustrations on the grassy turf.

From the time antler velvet is shed until the antlers fall months later, there is sporadic sparring among bulls, especially the young, aspiring individuals. Bouts may be lively and noisy but seldom result in serious injury.

the dead elk was a particularly vicious bull that had been engaged in many spectacular fights with other elk.

We have to wonder if Oregon elk are not exceedingly tough. During the 1982 deer hunting season, a seven-point bull had been shot illegally near Prineville. At a checking station, biologists found what seemed to be an antler tine protruding from his left eye. Len Mathiesen of Bend, Oregon, at first thought it was some kind of abnormal growth. But the tine turned out to be the main branch from the antler of another elk, a souvenir of a battle with that bull. The tine had been driven into the roof of the bull's mouth and out the eye socket, blinding the animal in that eye.

Amazingly the tine had been in that position for several years. After the fight, the base of the tine must have extended downward into the mouth where, over time, it became rounded and polished from wear, and stained the same color as the bull's teeth. According to Ken Durbin of the Oregon Wildlife bulletin, a large calloused growth had developed

on the elk's tongue where it had rubbed against the antler. Also the teeth were worn unevenly since the bull favored one side of his mouth while feeding. Still, there is more.

The bull in question was missing a large patch of hair from one hip, and an autopsy revealed large bruises in the muscle underneath the skin. Biologists had their probable explanation for the injury when a logger reported that he had struck a large bull blind-side with his truck. The bull had normal vision on only one side.

Longevity

Not all elk die early, violent deaths, and in fact a number of unusual longevity records are worth listing. Recently biologists examined the skull of a Jackson Hole elk known to be ten years old from the tag in its ear. The animal's teeth were in good condition and it could have survived longer. Rocky Mountain elk have lived as long as 18 years in zoos, and several Olympic elk managed to reach 14 years in captivity. I know of one bull elk on a northwestern game farm that, at this writing, is 13 years old and has sired between 60 and 70 calves.

Like any other wild creatures, though, elk have two different life-spans: the average and the potential, which are two greatly different things. Game managers in several western states have calculated the average life spans of their own herds. Wherever there is hunting, and herds are managed for the maximum harvest, four years is ripe old age for a bull and the average longevity is less than that. Average age for bulls is also less than that for cows where either sex can be hunted.

The true Methuselah of American wapiti, as far as we know, was one shot by an Arizona hunter in 1937. That one had been trapped and tagged 24 years earlier when it was already a year old. There is no record of the condition of that elk, the size of its rack (if any), or whether the hunter found the meat to be edible. It probably wasn't.

The Future

What is the future of the North American elk? Short-term, it is probably fairly good because a large elk herd is an economic asset to states such as Wyoming and Montana, where hunters spend a lot of money. For these areas elk are the geese that lay golden eggs. But it is doubtful that any of the great, big-game animals can long survive in more than token numbers if Americans continue to elect officials and tolerate agencies hostile to wilderness and the American environment.

DEVELOPMENT OF PUBLIC LANDS. Unfortunately, millions of acres of pristine lands are currently scheduled for some kind of development in every national forest in the country, as part of ongoing plans of the U.S. Forest Service. Thousands of miles of new roads will be built, involving clear-cutting of trees on hundreds of thousands of acres. Countless other acres will be sacrificed to new mineral exploration and development, while existing forest trails are slated for obliteration.

An example of this scheduled development involves the 24 roadless areas that remain in the Medicine Bow National Forest, Wyoming—a beautiful and popular outdoor tract. The Forest Service will allow logging in half the roadless areas. Oil, gas, and mineral development will be encouraged in the rest, driving out the elk and the people who like to see them. The Blue Range of Arizona's Apache National Forest, another unique haven of big game in that state, is also scheduled to be mostly plundered.

OVERGRAZING. Added to the problem of overdevelopment is the continuing overgrazing of livestock on public lands. The Forest Service, as well as the Bureau of Land Management (BLM), have for years been collecting less than half of what it costs to manage our precious rangelands. The investigative staff of the House Committee on Appropriations in Washington, D.C., recently reported that wildlife on BLM lands alone supported a recreational industry worth more than $241 million annually to the western economy. But too many cattle are gradually degrading that resource, and cattle owners are not beginning to pay for it.

ENCROACHING CIVILIZATION.   People have a habit of moving to the mountains to gaze at the scenery and enjoy the cool forests and wildlife. At the same time they are invading critical wildlife habitat, too often the winter range needed by elk. Sometimes when the wildlife presses too close, as in Estes Park, Colorado, where elk wander through backyards and trample gardens, people soon change their views about enjoying the critters. They want them removed. That is pretty much the situation all up and down Colorado's Front Range foothills, as just one example. At the posh Perry Park subdivision near Castle Rock, a cougar appeared to feed on the elk and deer that used to live there alone. But the cat soon found that catching household pets in backyards was easier, so it had to be shot by the Colorado Division of Wildlife.

Poaching

Poaching is another widespread problem seemingly on the increase. In the past, simply shooting bull elk to obtain antlers for the curio trade was serious enough. But antlers aren't just knife handles and cribbage boards anymore. Now there is mushrooming demand for antlers in the velvet to supply an insatiable oriental medicine market. Tons of elk antlers are illegally taken, ground up, and sold by the ounce as a supposed virility drug or to cure ulcers. Although some shed antlers are sold at public auction, the demand is now so high that prices have spawned an illicit trade in elk antlers satisfied only by poachers. With new highs in poaching technology come new lows in ethics.

Poachers have moved into such national parks as Yellowstone, and Kootenai in Canada, where the elk are less shy of humans. There they shoot bulls with tranquilizing darts, from guns that are quieter than regular big-game rifles. According to Hank Fabish, a warden for Montana's Department of Fish, Wildlife and Parks, the poachers only saw off the antlers. Few of these super-bulls ever recover because the poachers are untrained in tranquilizer dosages and in caring for the tranquilized animals. The carcasses remain as testimony to the illegal trade.

Recently an inflatable rubber raft full of antlers was discovered capsized on the Yellowstone River not too far from the Yellowstone Park headquarters. The boatman was missing and presumably drowned. Rangers wondered how long this clever method of carrying antlers undetected out of the National Park had been used before the accident.

Whether the problem is overdevelopment, poor land management, or poaching, the truth is that every time we do something else detrimental to our remaining elk herds, we are doing something worse in the long run to ourselves.

2.

MOOSE

November isn't always my favorite month. For one thing it marks the end of busy, happy, exhilarating autumn where we live. Wildflowers and weeds are dead or dying, and the wind is ominous, promising angry blizzards and snowdrifts. Fresh snow already shines on the mountains. The average low and comfortable humidity in northwestern Wyoming gradually creeps higher.

Still, there is something keen and exciting about November. A lot may be taking place in the newly damp, gray forest. The rut of the mule deer is one such thing. We also begin to see, for the first time since spring, a good many moose around our house. They come and go. It is almost as if they are prospecting for good places to spend the next few months, and for good food supplies when snow will be very deep. To live among moose and not admire them immensely is almost impossible.

In November, moose stand around and sometimes stare into my bedroom window. One bull liked to rub against the wooden posts supporting our back porch. Moose wander among the houses, which are not too close together on our Teton mountainside. Many of these houses are left vacant and cold by owners who only use them in summer. Moose stand on the sheltered sides in winter, out of the raw wind.

One thing I like about moose is that they are not intimidated, as are the mule deer and elk passing through, by house dogs yapping at their heels. Most of these dogs are wary enough to do their yapping a safe distance from heavy cloven hoofs. I've never seen an adult moose as much as turn and acknowledge that the dog was barking.

Moose Numbers and Range

Moose are doing fairly well almost wherever they live in North America today. Their numbers are very near the carrying capacity of their range completely across Canada, although nowhere are their population densities as great as those of other deer. Luckily, moose can cope with the excessive and often abusive timber cutting practices now condoned in both Canada and the United States. An intense competition exists be-

This Wyoming cow moose had ventured out onto a beaver dam and then into the water for new green growth.

PAGE 38
With antler velvet shedding in bloodstained tatters, this Alaskan bull will soon show the effects of his altered body chemistry during the fall rut. This will temporarily transform him into a lustful fellow—often frustrated and highly aggressive.

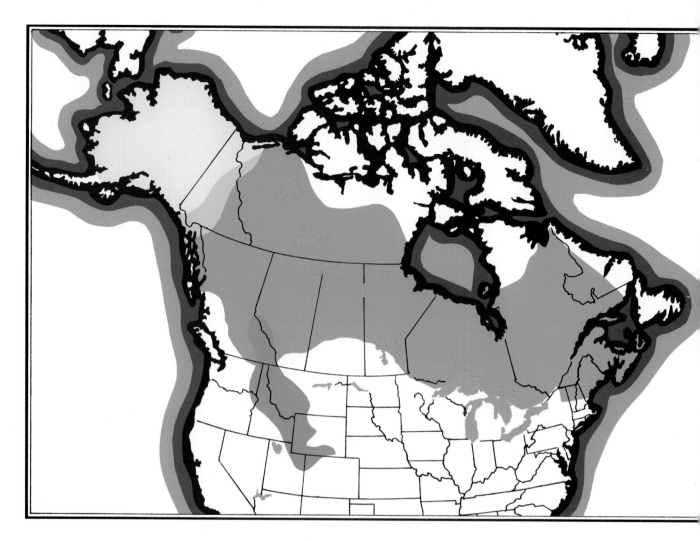

Here are ranges of the three North American moose subspecies: Alaska-Yukon, Canadian, and Shiras.

tween the two countries to cut and sell more and more timber, the consequences be damned.

Moose are now abundant enough in New England to hold an open hunting season in Maine. That may also happen in Michigan, where an estimated 1,000 live on the Upper Peninsula. Moose numbers are holding up well in Wyoming and Montana as well, and probably also in Alaska where they are an important subsistence animal for natives as well as a game species for hunters.

Moose are relative newcomers to the Yellowstone-Jackson Hole ecosystem. The trappers and explorers who vagabonded through Jackson Hole a century ago did not report seeing any. In 1880, only a few moose were recorded in Yellowstone by P.W. Norris, the Park superintendent. The Shiras subspecies did not become established in Jackson Hole until about 1910 when some animals wandered southward from Yellowstone, according to biologist Douglas B. Houston in his book, *The Shiras Moose in Jackson Hole*. By 1912, Forest Service census crews counted 47 in the area. But moose did not reach their current abundance until the 1950s when the population was estimated at about 600. Today the animals have become widely established through natural dispersal in Wyoming, where their numbers are estimated at 8,000. Grand Teton National Park is the best place to see them during any season.

The Physical Specimen

Moose are most often described as ugly, ungainly, or both. But beauty is in the eye of the beholder. I see the moose not only as the largest member of the deer family on earth, but also as a handsome, perfectly adapted survivor. That so-called ungainliness, incidentally, is deceptive. A moose can travel farther and faster through a marsh—where other heavy creatures flounder—or plow through deep snow that would soon enough trap the other animals covered in this book. Those long legs and long pistonlike strides are a powerful combination. On rare occasions when a moose becomes irritated, you are wise to stay out the way of the huge, pounding hoofs.

If you have ever watched and heard a bull moose lumber away through a dry forest, crushing debris underfoot and snapping off dead limbs of trees, do not be misled. Take my word that that same animal can also move silently, almost ghostlike, through the same timber, without scraping an antler on anything.

Now about that "ugly" visage. The long face and overhanging muzzle are no handicaps. Instead, they enable a moose to browse selectively on submerged vegetation while wading flank-deep in a beaver pond and at the same time either see or hear approaching danger. That same ample muzzle allows a moose to neatly strip the green leaves off thin willow shoots. Utilitarian rather than ugly is what the muzzle is.

The muzzle of a moose may not look handsome to many people, but it is well adapted for stripping the tender leaves and tips from willows and other browse trees as shown here.

PELAGE.   An adult's body is deep and heavy with long outer hair that varies in color from dark brown to black. The stiltlike legs are lighter, usually gray. An undercoat of fine wool beneath the dark body hair enables moose to withstand brutally cold winters by minimizing loss of body heat, while at the same time absorbing heat from the sun.

During a three-day blizzard in Jackson Hole, we saw a cow moose and calf bedded in snow 4 to 5 feet deep; neither animal moved more than 20 feet to change positions throughout that storm, which often covered them.

Later after another midwinter blizzard, Peggy and I were ski-touring on a trail from the White Grass Ranger Station to Phelps Lake, both in Grand Teton National Park. Moose often winter in this vicinity. The snow was very deep and the red trail markers, tacked to trees at eye level, were now barely above snow level. Because it was a cold day, we kept moving rapidly to stay warm. Suddenly coming around a bend where powder snow had blown and drifted, we came upon a moose that

arose out of nowhere squarely in our path. Peggy, who was leading, almost collided with it.

That moose may have been bedded in that spot since the beginning of the storm until it was almost completely covered over, except for its nose. Probably it was also fairly warm inside its snow cocoon.

SIZE AND WEIGHT. The moose we await every November is the Wyoming or Shiras subspecies, one of seven that occupy northern conifer forests around the globe from Eurasia to North America. In northern Europe a moose is called an elk or *elch*, the name that early American settlers mistakenly applied to the North American wapiti, now known as elk. To a lay person, all of the subspecies will look alike and be indistinguishable. However, the Wyoming moose is among the smallest, while the Alaskan, or Alaska-Yukon, race is the largest. A full-grown Wyoming male could reach 1,200 pounds. A healthy Alaskan bull standing 7 feet at the shoulder might exceed a ton.

Now in summer, moose are rounded and healthy again, dining on favorite foods that are seasonally abundant. All moose experience cycles of feast and famine.

PAGE 45, TOP
The time is spring, and the antlers are still in velvet and only partially grown.

PAGE 45, BOTTOM
It is the end of the short Alaskan summer, and this fine bull moose sheds the velvet from his antlers as he feels the first effects of the oncoming rut. Now is a dramatic time to be photographing formidable animals such as these.

PAGE 46
This is an extremely large, massive specimen of the Shiras; it approaches record dimensions. Yet antlers of the Wyoming, or Shiras, moose are neither as heavy nor as splendid as those of the Alaska-Yukon animals.

PAGE 47, TOP
Moose are often said to be solitary animals, but that is not so. Here are three Shiras bulls together on a cold November morning. In this same vicinity on Thanksgiving Day, I counted 33 moose in sight at once.

PAGE 47, BOTTOM
Peggy and I found this moose skull and antlers beside the American River near Katmai National Park, Alaska. The bull may have been killed and eaten by wolves.

Antlers

Commensurate with their bulk, Alaskan bull moose also sport the largest antlers grown by any living animals. Some palmated racks span more than 6 feet and weigh more than 75 pounds. It is almost inconceivable that such a mass of calcium can be grown in a single four-month season, as well as year after year.

The antlers sprout from pedicels on opposite sides of the skull in April or early May. From then until late August or September, the speed of the growth exceeds any other such growth in Nature. Throughout summer, as with the other deer, the relatively soft, blood-engorged antlers are covered with a velvety coating. This falls away or is rubbed off in early September, sometimes slightly sooner, both in Wyoming or Alaska.

PRIMARY FUNCTION OF ANTLERS. Biologists debate whether the bull's antlers are decorative—to impress and intimidate—whether the primary function is as a weapon, or both. While it is true that an especially heavily antlered bull needs only to flaunt his antlers to establish dominance in a rutting area, bulls with nearly equal-size racks will join combat to settle the matter. In any case, the winner gets to do most of the breeding, which allows the strongest, most vigorous males to pass along their genes. A good argument for the antlers-as-weaponry theory is that following the rut, when moose no longer need the head-gear, they shed it.

VELVET RUBS. There is also disagreement over whether bulls deliberately rub their antlers on trees and brush to get rid of the velvet, or whether the dried, loosening velvet just happens to be brushed off when they polish antlers prior to the rut. I lean toward the latter. Too often, both in Yellowstone Park and Denali National Park, Alaska, I have watched large bulls tolerating the masses of damp, darkened velvet hanging down in their faces for days until it fell away naturally. The animals didn't even make a pass at a bush. On the other hand we have also photographed bulls that seem to be polishing antlers at the same time they really tried to remove old velvet.

STRONG NECKS. Carrying such awesome symmetrical antlers develops strong necks, which are important to bulls in the combat that takes place during the breeding season. But what seems most remarkable to me is illustrated by a post-rut Alaskan bull I saw carrying just the right side of his rack. The left had recently fallen off, judging from the bloody pedicel. Even though the half antler must have weighed 30 to 35 pounds and extended 3 feet out to one side, the bull still walked and browsed with his head straight up in the normal position.

SHED ANTLERS AS FOOD. Even the dropped antlers are part of Nature's endless cycle. Voles, porcupines, Arctic and related ground squirrels gnaw on the antlers for the high calcium content. More than once (in Wyoming) mule deer have been reported nibbling on discarded moose antlers.

Moose Stories

Mostly because of a moose's impressive size and apparent truculence, almost everyone who lives in moose country has moose stories to tell—some real, some imagined, and many a combination of the two. For example, moose will invade barnyards during the depths of winter and kick cattle away from the hay supply, but not as often as claimed. Odd ones will also stroll through small rural towns, stopping traffic and challenging yellow school buses for the right of way. In the winter of 1984 a moose cow alternately munched on ornamental shrubbery and slept in a snowbank downtown beside the Teton County Courthouse in Jackson Hole. Hundreds of people photographed the animal from a few feet away without any problems.

But Maine game warden Dave Priest is one person who had a problem. On patrol one day in the Soubunge Mountain region, he came upon a bull ensnared in fallen telephone wires. The bull may have pushed over a telephone pole; now the animal seemed hopelessly wrapped in wire by its jaws, his legs, and antlers. Priest took aim with his axe to try to cut the restricting wire.

His trouble started because his aim was too good. Wire snapped, and the moose lunged and was nearly free. In the same split second, the moose was on his hind legs lunging at his savior. The bull struck Priest on the shoulder and almost scored a knockout, but the wires still around the animal's legs prevented him from slamming and pinning the game warden to the ground. Down on all fours, Priest managed to escape around the back side of a beech tree. The last he saw of the bull, the animal was trotting away with a nest of telephone wire still tangled in his rack.

Another interesting moose story was related to me by a backcountry ranger in Yellowstone National Park. For about a month one summer, an old bull moose lived near the remote ranger station, and was seen by the ranger almost every day. The strange thing is that the moose was always accompanied by a large bull bison. Whenever the ranger tried to approach the moose to take its picture, the other half of the odd couple would charge him. One day the two unaccountably vanished from the area and were not seen again.

One day Gene Wade, a retired pack-trip outfitter of Cooke City, Montana, was leading a trail ride in the Specimen Ridge section of Yellowstone. Not far from an overnight camp he lost the trail and found himself riding through an unfamiliar and lonely basin. What caught his eye at first seemed to be a collection of antlers hidden in tall grass. His first thought was, Who has taken the trouble to pile up antlers way out here?

On riding closer, Wade saw that it was not a man-made pile, but rather the complete skeletons of two bull moose with antlers locked. It seemed that one had thrown the other over his shoulder, apparently breaking the other's neck, and both animals died in that position. Coyotes and probably ravens had eaten away the flesh, leaving the skeletons largely intact. Wade reasoned that the fight had taken place after grizzly bears went into hibernation, because when bears find such a bonanza they are likely to tear it apart when eating it.

Diet

Moose of the northern Rockies prefer willows, and plenty of them, especially Geyer's and blueberry willows for year-round sustenance. No other kind of browse is nearly so important to them. Because the most lush willow stands grow along rivers and in wetlands, I usually associate moose with such places. Consider how many photographs are published every year of moose wading in ponds and streams, where they eat a variety of aquatic vegetation including water crowfoot, hornwort, and leafy pondweed.

Yet in summer we often meet moose far from familiar moist haunts, in distant high alpine areas near timberline, along the lofty Teton Crest Trail, for instance. One explanation might be to avoid the troublesome insects at lower elevations. But the animals might also be varying their diets with mountain grasses, forbs, chokecherry, serviceberry, quaking aspen, snowbush, and red osier dogwood, all of which they seem to relish periodically. In any event, a moose's ability to browse higher than other animals is an advantage, particularly in winter. It means that they do not have to travel far to avoid really deep snows as other big-game species must.

Often when hiking or cross-country skiing around our home, we have come upon areas—usually in a spruce or Douglas fir woods—where the ground was virtually layered with the unique, hard oval pel-

Mute evidence of the hazards of wearing a monarch's crown, these death-locked combatants are returning full measure to the ecosystem that sustained them.

let-droppings of moose. What we found were wintering areas where moose might remain for weeks, maybe months, moving very little until the season passed. Some moose may prefer subalpine fir to willow as the most important winter food.

Once in a high spike camp on Mt. Assiniboine, British Columbia, a moose loitered nearby our tents for two days, and at night licked the cold ashes of our campfire. One member of our party was city-bred, French, newly arrived in Canada, and on her first trip into a wilderness. One night the moose tripped and became entangled in her tent ropes and then panicked, dragging the whole tent down a steep mountainside before the lady could escape. She didn't sleep a wink the rest of the trip, and she vowed never to set foot in any moose wilds ever again.

Gregarious Behaviors

Of all our big-game animals, moose tend to be the easiest to observe and study. They're big enough and dark enough to be easily spotted, and any creature that size has a hard time hiding. In addition, moose are relatively unwary. Given some protection, as in parks and other sanctuaries, they're even fairly trusting of people. So why is it that moose watchers, many writers, and even a few biologists (who should know better) keep repeating the same tired old myth that moose are a solitary species?

True, moose are more solitary than elk and caribou, or even mule deer, which are definitely herd animals. But if you spend much time

It is April, and this thin cow has just managed to survive the brutal northwestern Wyoming winter. She is shedding her old coat, but may soon be able to find enough new green vegetation to regain strength and grow sleek again. She almost certainly carries a fetus, which will be dropped in a month or so.

PAGE 50
A young Yellowstone moose pauses in midstream to scan the July landscape. And no wonder—a grizzly bear with twin cubs is crossing the green meadow not far away.

A Wyoming female in estrus is invariably followed closely by a bull, or several of them. I have seen as many as eight suitors in a following procession. The heaviest and most powerful of the bulls do most of the breeding.

throughout the year in moose country, you will find two or more together more often than one moose alone.

SOCIALIZING DURING THE RUT.   There is that brief period in early springtime just after a cow chases away her previous year's calf or calves (not always successfully) and before the next ones are born when a cow moose may be solitary. Other than that, cow moose are always accompanied. Even during the rut and when in estrus, a cow is likely to be trailed not only by several suitors, but also by her calves. In October 1982, near Moran, Wyoming, I counted eleven moose in a group that included a very popular female, twin two-thirds-grown calves, and eight bulls of various sizes. Such congregations are not at all unusual with Shiras moose throughout the rut, which may wax and wane for a month or more. In fact, in a single morning's drive, we have counted as many as 33 moose, mostly bulls, associating together in the same Moran meadows.

Now let's consider winter, a season when, as I mentioned before, moose venture into our front yard and vicinity without being baited and without enticement of any kind. Rarely do we see an animal alone. If male, there are usually two together, and we once had a group of five make an extended visit. I could go on and on, quoting from my notes, but the plain fact is that Wyoming moose at least are far more gregarious than most people realize. Maybe they just look as if they should live alone.

ALASKAN MOOSE.   Peggy and I have not spent nearly as much time among Alaskan moose, but I do not believe they are much different from Shiras moose. Drive or hike through the moose range of Denali National Park, and the animals you see will be cows with calves or they will be a couple of bachelor bulls together. Perhaps the largest of the bulls will summer alone, but as August blends into golden September they will move toward such traditional rutting areas as on Igloo Flats.

Courtship

As August comes to an end, the summer tolerance among bulls begins to disintegrate. Some larger bulls may go separate ways, testing one another whenever they meet and polishing antlers, which become fairly large and palm-shaped by their third or fourth fall. By the end of September a hiker on northwestern Wyoming trails is likely to encounter a bull or bulls anywhere, walking aimlessly it may seem, always grunting, in a search for receptive cows. Such cows may not be hard to find, but access by a wandering bull is another thing altogether.

The next event, on meeting, is the moose courtship ritual, which may last a week or more and is similar to the courtship of some bird species. Because the aura of estrus must be detectable for miles carried on the autumn wind, a bull must establish dominance over other bulls almost certain to be vying for the same cow. After that the bull stands posing interminably, often for days, always sideways and within a short distance of the cow. The cows we have watched paid no noticeable attention to the display and kept feeding, while the bull followed a step or so behind, as if in lock-step. Gradually the eager bull, still grunting, becomes more persistent and eventually, if a still bigger bull does not show up, breeding takes place. If a larger or equally capable bull does arrive, a really savage, lunging fight occurs and injuries may be inflicted.

Two young Alaskan bulls face off before a shoving match during the early stages of the fall rut. I first heard the animals sparring, antlers crashing, from several hundred yards away.

A trio of Alaskan bulls square off in a three-way match near Savage River, Denali National Park. On several occasions I have seen a third animal join a dueling pair in a contest of strength.

Fighting

Serious bull fights tend to be head-to-head pushing matches with antlers. The fights are settled by power and determination rather than a swift and slashing attack. But they normally do not last long. One male may soon realize it is the loser and pull away. If it stumbles or falls, however, its flank may be bruised or punctured by the rival's antlers. Being the loser doesn't necessarily mean leaving the scene. As long as the loser keeps a certain distance, the winner will probably tolerate him and other lesser bulls in the vicinity. Chasing them farther away may not be worth the effort and risk.

Twice Peggy and I have seen third bulls join in a fight between two others, though we cannot understand it. Once we could not see the conclusion, and the other time the animals broke and separated. But later we again heard the rattle of antlers farther away.

The Rut

Igloo Flats, Alaska, is a wonderful place for a moose watcher or photographer to be around Labor Day. For that's about when the rut begins. Bulls of various sizes and credentials arrive punctually in the vicinity. The early days are spent in shedding the velvet of summer, in dueling first against brush or sapling opponents, and finally in testing one another. The arrival of a cow soon coming into estrus, followed by another, stimulates the competition and the action even more. At this time several super bulls also make their appearance. The only trouble for photographers is that this is also a season of cold drizzles and early, wet snowfalls. You need an ample supply of equanimity and patience, plus reliable foul weather gear, to catch the Igloo Flats happening on film. But it is surely worth trying.

THE ESTRUS CYCLE.   Estrus in cows lasts for less than 24 hours. But if the female is not bred, estrus recurs one or two times until about November. Biologists believe that in Wyoming the most successful estrus cycle is the second, usually in late September or early October. When one breeding is over, the male departs and searches for other action elsewhere as his strength and stamina permit.

Both males and females only 1½ years old are physiologically capable of breeding but would do so only in unusual circumstances. Cows normally have their first calves at 2½ or 3½ years, and can continue to do so as long as they live and stay healthy, possibly as long as a dozen years.

Births

After eight months' gestation, moose cows retreat to secluded places and give birth to a single calf and occasionally to twins. A cow near Wonder Lake in Alaska's Denali National Park was observed with triplets and may have dropped three calves two seasons in a row. In Wyoming the calves arrive in late May and June. The newborn are extremely appealing, helpless, grotesque, or all of these, depending on your point of view. The first impression is usually that a calf is all legs. But not many days pass before it is able to closely follow its mother anywhere.

DO COWS PROTECT THEIR YOUNG?   To say that a strong bond develops between a cow and her calf is an understatement. Moose may not be protective of space or territory, as are some other large

It is autumn just before freeze-up, and moose are still visiting ponds to browse on the aquatic plants. The cow and calf visited this pond in early morning.

PAGES 56 & 57
Wonder Lake, Denali National Park. Peggy and I sit near the water's edge at dusk awaiting the moose that will emerge and browse on the aquatic vegetation. First to arrive on this haunting wilderness scene were this cow and calf. It was a most memorable evening.

mammals, but they are most protective of their young. We always give cows with calves, especially very young calves, a wider berth even than bulls during the rut. If you see a cow mother with ears laid back and shoulder mane erect, it is time for you to be halfway up a tree.

Summer is a season of plenty for cows, a time of accumulating fat in a familiar territory. A typical scene is one of the not-so-solitary female, now sleek and looking better every day. She has a small calf underfoot, while another calf nearly her own size trails just behind—especially if it is also a female. The males, also in small groups, wander farther during the summer months; ear-tagged males have traveled as far as 40 miles from their home range. These bulls also accumulate a lot of weight and strength, which they will soon need for yet another rut.

Collision Course with Man

Probably due to the increasing number of people settling in Alaska, and the resulting pressure on the animals, moose are often in the Alaska news. An angry moose forced veteran musher Susan Butcher out of the annual Idatarod Sled Dog Race. This moose was just one of scores killed in the Susitna Valley during just three days in March. The others were ground into mooseburger by trains running between Anchorage and Fairbanks, a region that is good moose habitat.

Butcher and her 17-dog team were leading the pack of 54 mushers on the second day of the grueling 1,130-mile trail to Nome. She was a favorite, traveling on a fast trail during a cold moonlit night. On suddenly rounding a corner, the dogs overran a moose and bedlam followed. The cow killed one dog and injured seven others before another racer appeared and killed the flailing moose with four shots from a handgun.

The night before Butcher had to drop from the race, Alaska Railroad trains killed 21 moose in the 30-mile stretch between Willow and Talkeetna. Three days later, the toll was 68. The Railroad keeps no record of moose kills, but the Alaska Fish and Game Department reports a minimum annual average of 50 dead. The toll was 171 in 1978, and 115 in 1970, both heavy snowfall years. In 1979 state game biologists, walking the tracks in early May, counted 200 moose carcasses between Knik River and Talkeetna. The huge animals are trapped in the deep canyons of snow along the railroad right-of-way and, of course, are unable to outrun or evade the speeding iron predators.

On March 13, 1985, with the number of moose killed that winter at 165 and rising, one moose got even with the railroad—posthumously. It derailed an engine and five coal cars, closing the crossing at Talkeetna for four hours. Some townspeople cheered the event.

Obviously there is a desperate need for some device, some technique, some sure-fire method that will keep moose off the tracks. Alaska Fish and Game Department personnel have tried everything from ultrasonic sound waves—which supposedly only moose can hear—to blinking lights, including the elimination of ditches on either side of the tracks. The ultra-sonic gadget seemed to work at first, but was ultimately unsuccessful. In addition, the train speed in the critical area where about

7,000 moose winter was cut to 30 mph, and the high-beam headlights that seemed to hypnotize moose were dimmed. But the carnage goes on. So, too, does the search for a solution.

Live-Trapping Methods

Because of their size, shape, and strength, moose have tested biologists and game managers in still other ways. To manage them wisely means that it is necessary to live-trap, examine, and tag some of them so that their movements can be monitored. That is not too difficult in winter areas where moose can be enticed into traps with alfalfa, grain pellets, and salt blocks, and then tranquilized and collared. But it is an entirely different matter when the animals live in a roadless wilderness that is half land and half water and difficult to traverse at any time, as in vast tracts of eastern Canada. The only practical solution, though expensive, is to use a helicopter.

HELICOPTER TAG.   Pursuing and then handling mammals that can weigh a ton, by helicopter, isn't an easy way to make a living. There are no dull moments. In Ontario, the job is done in late June or early July, during mornings and evenings when the moose are frequenting shallow lakes and marshes to feed on eelgrass and yellow pond lily. Pilot and crew cruise at 800 to 1,000 feet, watching for animals in or near the water. When a moose is spotted, the pilot shifts into auto-rotation, by which the craft descends very rapidly, hovers, and drives the target moose into deeper water where it must swim.

Now for the ticklish part. Just before the pilot touches down, a biologist-tagger crawls out onto the starboard pontoon and lies there prone, reaching as far forward as possible. The pilot then taxis the chopper so that the pontoons straddle the swimming moose while the biologist affixes a metal tag and/or colored streamer—for distant identification—with a pair of livestock pliers.

The whole task may require as long as 45 seconds, or as little as eight (a record) by an experienced, coordinated team. In one very busy day an Ontario crew was able to find and tag 50 moose during an actual flying time of 10 hours and 35 minutes, or nearly five moose per hour.

A Famous Moose

Except for killer bears, not many individual big-game animals ever achieve even a small measure of fame. One exception was a bull moose known as the Missouri Kid, an extraordinary animal and the Marco Polo of all moosedom. His travels were chronicled by feature articles in *Sports Illustrated* magazine and most newspapers of mid-America.

We know that the animal was probably born in spring 1975, somewhere around the International Boundary between Lake of the Woods and Lake Superior. But sometime in the fall of 1976, he began legging his way southward through Minnesota, the exact route unknown. In December 1976, the moose—now wearing antlers—showed up near

the village of Emmetsburg on the Des Moines River in northern Iowa. He settled down there to the delight of natives. Going that far was an incredible journey for a moose.

The Missouri Kid remained near Emmetsburg until the fall of 1977 when he suddenly abandoned the many human friends he'd made there. Dogs may have driven him away. Or the search for female companionship. By late October, the moose was seen near Boone, Iowa, and soon passed through the Des Moines area where a quarter of a million people live among a maze of concrete highway interchanges. Before Christmas he crossed the Iowa line into Clark County, Missouri.

During the first half of 1978, this globe-trotting moose was seen only occasionally in northeastern Missouri. But after that, more and more people noticed the bull touring the Show Me State, negotiating roads, swamps, towns, farms, and shopping centers, crossing rivers, and looking around. The Missouri Kid often appeared on evening television news in some new locale. The Kid was last seen in the flesh in February 1979 near Louisiana, Missouri, near the Mississippi River— after which he vanished.

Marion Trayner was scouting the woods near Bowling Green, Missouri, one day for a place to hunt turkeys when the spring gobbling season opened. By accident he found a pair of shed moose antlers, which, allowing for skull space, would have spread 41 inches. Certainly these had belonged to the Missouri Kid, which some speculated may have died trying to swim the Mississippi. The animal could also have been poached, he might have returned to Minnesota (which is unlikely), or he could have lived out his life unnoticed in a dark Missouri riverbottom (which is even more unlikely).

But how can anybody explain this three-year odyssey of at least 1,500 miles by an animal that normally spends a whole lifetime in a very restricted range?

A Friend to Moose

A moose may have a face only a mother could love, and standing next to a whitetail deer moose may seem unattractive, but they have a special appeal. A rural Quebec housewife who resented the squirrels invading her bird feeders did not mind when a cow and calf ate her entire vegetable garden. In Ontario a farmer shot a neighbor's poodle that was annoying "his" moose. Moose also found a special friend in 71-year-old Gail Frazier of Orofino, Idaho.

In the late 1970s, foresters planned to build new access roads for logging in the Elk Summit area of the Clearwater National Forest. But conservationists halted the road building and the timbering on the grounds that this was an area vital to moose survival there. Biologist Dan Davis had begun several studies of the moose, radio-collaring some and dye-marking others to keep track of them. But as always in recent times, funding for the animals always ran out. Now meet Gail Frazier.

After 20 years as a checkout clerk in a supermarket, Gail lived in a modest camper trailer and still worked part-time. But the clerking was unfulfilling and she volunteered to monitor the moose, which she had heard about. To do so she had to tow her camper 150 miles away from home into a remote area. There, with only a Norwegian wolfhound named Rufus for company, she makes meticulous daily observations of from 15 to 50 moose to see how they live and where they wander. Many have become her friends, with such names as King Solomon, Greta, Bullwinkle, and Victoria.

Gail soon learned that ear tags and colored streamers were not necessary to identify individual moose. The animals have flaps of loose skin called bells hanging from their throats, and no two sets of bells—or antlers, for that matter—are alike. With little more than a glance, she soon could tell Greta from Victoria or any other moose. If the moose eventually make it on Elk Summit, they can thank this dedicated woman.

Moose and all wild creatures need more friends like Gail Frazier.

3.
CARIBOU

The eerie and treeless tundra cuts like an irregular swath around the Arctic, including much of northern Canada and Alaska. It also curves across Greenland into northern Europe and continues across Asia, covering about three million square miles, or roughly 5 percent of the earth's land surface.

Although the summer sun's heat can be oppressive in the tundra regions, the same sun appears only briefly on the southern horizon in winter. July and August temperatures of 80° F or higher plunge to minus 70° or lower. That extreme temperature range is unmatched even in Antarctica, and life in such a climate zone can be extremely harsh. But every kind of country is good for something.

Little rain or snow falls here—little more, in fact, than in many of the world's desert areas. But just beneath the surface of this lonely land are immense amounts of water, frozen as ice for most of every year in an earth layer called permafrost.

From mid-September until about mid-June, the tundra of North America is a white, cold, apparently sterile and lifeless land. But then a miracle takes place. Suddenly in June, hundreds of low-growing plant species are engendered by a sun that now only briefly sinks below the horizon. Wildflowers in astonishing variety and color spring up and, for a short time, make the landscape as colorful as any place on earth. Countless waterfowl and shorebirds, some of which have flown from as far away as the Antarctic islands, descend on the tundra. Quickly they mate, nest, raise young, and depart for the south again before every pond and trickle freezes up.

But even in that abbreviated summer, all is not unadulterated warmth and beauty. The steady, terrible whine of mosquitos drowns out the lower-keyed hum of other insects. On a windless day there is no escape from them.

Lemmings race over the tundra in search of food, always wary of the snowy owls, Arctic foxes, wolves, and long-tailed jaegers that are looking for them. Stand quietly and you may also hear the cackling of ptarmigan above the cacophony of geese and loons and sandhill cranes.

Arctic wolves are the principal—and often sole—predators of caribou.

PAGE 62
A barren-ground caribou is a handsome animal, well suited to its environment. Its antlers are larger in relation to body size than those of any other American deer.

During summer, scattered bands of caribou wander over lonely uninhabited wilderness landscapes such as this in Denali National Park. This is the only American national park in which caribou are easy to find.

MAP, PAGE 65
Here are ranges of caribou for trophy classification recognized by the Boone and Crockett Club: barren-ground caribou (Alaska, Northwest Territories, and northern Yukon); woodland caribou (Eastern Canada); and mountain caribou (Alberta, British Columbia, and also the southern Yukon).

Especially around the southern edges of the tundra, where fingers of boreal forest begin, you may spot a moose or a barren-ground grizzly with cubs.

But here the most striking living creatures are the herds of caribou, which even at summer midnight cast long shadows before they become dark silhouettes against the orange sky. Caribou are the deer of the north. They are completely at home in the tundra, which in most months might be called the country of winter.

Peggy and I regret that we have spent so little time watching and photographing the caribou, as compared to other North American antlered animals. Yet caribou country is too difficult and too expensive for us to reach often from our home in western Wyoming. But anyone observing caribou quickly realizes that these sometimes unpredictable mammals are the central creatures and the pulse of life in the far North. They give life and movement to that remote region. Wolves, foxes, ravens, and a good many humans depend on them. Unlike the nearly vanished bison, caribou travel over the tundra today much as they always have.

Range and Numbers

In their trophy records program, the Boone and Crockett Club separates North American caribou into three categories: woodland (roughly Newfoundland and eastern Canada), mountain (Alberta and British Columbia), and barren-ground (Alaska, Yukon, and Northwest Territory).

Scientists are more likely to divide caribou into three separate populations. The Ungava population contains an estimated 150,000 animals living on 100,000 square miles of range in Canada east of Hudson Bay. The barren-ground population has contained from 500,000 to 600,000 animals living on the 600,000 square miles between Hudson Bay west to the Mackenzie River. The Alaskan population also contains about 500,000 to 600,000 caribou living on 400,000 square miles west of the Mackenzie River. Caribou no longer inhabit former range in New Brunswick, Nova Scotia, Maine, and Minnesota. Except for Alaska, the only caribou left in the United States are the 25 or so still clinging to existence in extreme northern Idaho.

While the population figures above may seem large, they are only a fraction of earlier numbers. In 1935, biologist Olaus Murie calculated that there were between one and two million caribou in Alaska. Wildlife biologist C. H. D. Clarke estimated that 3.8 million caribou lived in all of Canada in 1940.

It is easy for an amazed, untrained observer to overestimate the number of caribou in a given geographical area. During a major migration it is possible to stand in one place and watch an endless stream of animals flowing past from horizon to horizon, as I did once below Polychrome Pass in what was then McKinley National Park. But keep in mind that at such times all the caribou of a vast area may be in a single

Peary's caribou (classed as a barren-ground for trophy purposes) is the northern-most, cream-colored race named for Arctic explorer Admiral Richard Peary.

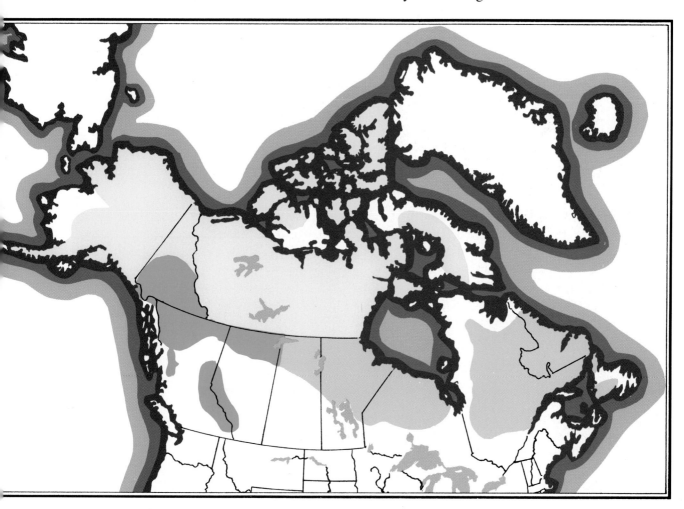

herd passing before your eyes. The other area you cannot see all around you may be empty of the animals.

Each of the three caribou populations is further subdivided into herds. For example, there are seven herds in the Alaskan population. Those are the Western Arctic, Central Arctic, Porcupine, Fortymile, Delta, McKinley, and Nelchina herds. Each occupies and travels within its own fairly well defined geographical area.

The Spring Migration

A caribou's life is one of constant movement. But no journey is more important than the spring migration that takes the herd from its wintering area to distant mountains or to tundra where a new generation will be born. It is rarely an easy journey because the cows may have to carry their calves for hundreds of miles, while tracked by wolves through slush and screaming storms and across swollen streams choked with ice—all without veering from a path that may be centuries old.

It is probably wrong to call this a spring migration, because spring in the Arctic is definite only on human calendars. The Porcupine, which is an international herd, moves back and forth between Alaska and Yukon Territory. The herd begins the trek in late March or April when even the larger rivers are still frozen, although not thick enough for safe passage of a large herd. Even in April the mercury can nosedive to 40° below. But that doesn't stop the herd.

Northern caribou are as mysterious as they are migratory. For ex-

In the central Canadian Arctic, caribou herds must often travel over pothole country such as this between summer and wintering areas. Their hoofs are well designed for such difficult terrain at any season.

ample, biologists are baffled by what triggers the migration. Changes in snow cover and increasing daylight must have some effect and so do the physiological changes taking place in the females. But scientists still wonder why there is as much as a month's variation in the exodus from year to year, and why one herd will start moving in one direction weeks after the nearest herd begins to move in another.

Spring migrations are led by females, probably by those most anxious to reach the calving area soonest. Most calves of the previous spring follow their mothers, and this may be the way in which they learn the migration route. Some of the one-year-old male calves may drop out of the migration, never really traveling all the way to the calving site. Mature males rarely accompany the pregnant females, preferring to dawdle on the winter ranges until more vegetation is exposed by melting snows. They wander later at a more leisurely, majestic pace, all the while growing the splendid massive antlers they will need later on. The breeding males will not catch up to or see the cows again until mid- to late summer.

George Calef, a Canadian biologist who has devoted much of his busy career to studying caribou—and authored the splendid book, *Caribou and the Barren Lands*—has noticed some other interesting facts about the spring migration. If the trip begins late, the herd must travel much faster than normal, covering 15 to 20 miles a day so that the pregnant cows are on the calving grounds when calves begin to arrive. Forty-mile dashes to arrive in time are not uncommon.

Calef has also learned that over the millennia caribou have learned to be efficient and practical overland travelers. They seek routes of least

Bachelor bulls and small bachelor herds seem to wander aimlessly during the summer, but their unpredictable treks always take them to suitable winter range.

resistance—surfaces of hard frozen lakes, places where snowpack is thin or blown away—always in the tracks that herd leaders have found to be hardest packed and best. Moving in single file, thousands of animals will use the exact tracks of the one ahead, compacting the snow to total solidity. In this way, the herd conserves a great amount of energy that would be lost if each caribou made its own tracks.

The ingrained instinct to find the easiest trails is so great that caribou will not hesitate to follow winter vehicle roads and even the paths of snowmobiles. Biologists have deliberately made such easy trails to direct traveling animals into traps where they could be examined and marked. Calef once had a caribou follow his own snowshoe trail into his camp. But most wild caribou have no human tracks to follow, and the spring migration remains a grueling, testing experience.

Physical Characteristics

Except for Peary's caribou of the northernmost Canadian Arctic islands, which are pale-gray or cream-colored compared to the darker gray of other caribou, the six subspecies of *Rangifer tarandus* differ little from one another.

The unceasing cold eight months of the year that makes survival difficult or impossible for less hardy animals is no great challenge to the caribou. The coat that covers the animal's entire body insulates so well that even during the lowest temperatures caribou never shiver. Little or no increase in metabolism is necessary to maintain body temperature.

We watched this barren-ground Alaskan bull pause every few steps to scratch with a hoof behind his ear, probably to dislodge a tick or other parasite. Insects are a nuisance to caribou throughout the summer.

The handsome, long-haired coat features long brittle guard hairs that taper in both directions from the middle forming a seal that protects the thick, fine wool underneath. These cellular guard hairs also help the underfur seal warm air close to a caribou's body. Except for the nostrils, even the muzzle is furred so the animal is insulated there when it probes into snow for food.

CIRCULATORY SYSTEM.   A super-warm coat alone would not be enough to suit this northern deer for life in a relentlessly cold climate. According to George Calef, caribou have what is called a counter-circulatory system in their legs similar to one found in certain Arctic birds and waterfowl, in the flippers of ocean mammals, and in a few other creatures restricted to the coldest latitudes. In the caribou's circulatory system, blood vessels from the legs to the heart run very near to those flowing in the opposite direction, permitting an exchange of heat from warm blood to cold blood. The system helps maintain the temperature in a caribou's legs at about 50° F, or well above freezing. The normal body temperature holds at about 105° F.

HOOFS.   Many of the various Indian names for caribou translate into "the digger" or "the pawer," and that is no wonder. The same species that might be conceived, born, and killed in snow also spends the greatest part of its life digging in snow. Not surprisingly, caribou hoofs have evolved into very efficient tools for the purpose. The shape of a hoof, which in an adult male measures 5-by-5 inches, also changes to adapt to the seasons. Walking on solid ground in summer rounds off the edges. But after walking in the first fall snows, the hard outer shell of each hoof—no longer eroded and honed by hard ground—develops a sharp edge and therefore a concave shape. Now the hoof is more suitable for scooping out snow, breaking through an ice crust, and getting better traction on ice.

ANTLERS—MALE AND FEMALE.   There are few more splendid sights in the North American wilderness than that of a mature bull caribou, white mane blowing in the wind as he stands on an amber mountainside displaying scarlet antlers against a bright blue or threatening black sky. This spectacle has given more than one photographer a case of buck fever, or worse, causing a botched exposure or focus.

Caribou are unique among North American deer in that females as well as males have antlers. Only in eastern Canada, in the Ungava population, do many of the females remain antlerless. A bull's antlers may grow to well over 3 feet high, with long flattened or palmate beams containing as many as 40 points. One, or rarely two, palmate brow tines or shovels extend straight out over the face. It is easy to tell a cow from a bull at any time by the comparatively smaller size and mass of the cow's antlers. Bulls are also much deeper and heavier in the bodies.

Shedding of velvet from antlers begins in late August and requires a few days to fall completely or be scraped away. At first the bare antlers may be covered with blood from residual circulation beneath the velvet, but that soon dries and blows away. I have found small trees and shrubs

that bulls have rubbed and twisted, but I do not believe that caribou spar with trees as much as the other antlered species. The oldest, largest bulls lose velvet first, followed by younger bulls; cows retain velvet until after the rut and use their antlers after calving to ward off competition for food.

Feeding Habits

Probably more than most other horned and antlered animals, caribou adapt their lives and activities to the weather. Falling mercury precipitates more activity, most of it involved in feeding. The extra exertion of searching and digging for food, as well as fermentation of fresh food in the stomach, produces warmth. When the mercury continues to drop and windchill increases on the barrens, whole herds of caribou can be found lying down, curled up to save heat loss. Individual animals may also dig out beds in the snow for protection and remain motionless as swirling snow partially covers them.

During a typical winter, snow can be a greater sapper of strength than bitter cold. Deep snow makes it harder to escape predators and more difficult to obtain food, which they locate by smell. Traveling along, the animals thrust their muzzles into the snow in search of edible plants. One year when early heavy snowfalls in Denali National Park had hastened the arrival of winter in the Alaska Range, I watched a herd of caribou migrating past Wonder Lake. The animals moved along steadily and seemed to locate food unerringly. They never stopped to dig without finding something; indeed, they were able to smell the food in soft snow that was then more than a foot deep everywhere.

In contrast to some more sedentary animals—particularly most domestic animals and even the closely related European reindeer, introduced into Alaska from Scandinavia—caribou keep feeding and moving in a manner that does not overgraze their range. When desirable foods in an area have been consumed, they push on. Herds normally feed at daybreak and at dusk, resting and ruminating during the middle of the day. In winter they also forage once at night. In the far North, all winter feeding periods may be in almost total darkness. Because food is harder to find and of poorer quality in winter, a caribou's growth ceases from early November until about the following May. Reserves of fat in the body cavity, along the saddle, and in the bone marrow are gradually used up. Over winter, each animal loses about 10 percent of its body weight and perhaps as much as 20 percent when the winter has been severe. It is a thin and an emaciated herd that emerges from its snowy winter area and begins drifting, shadowlike, on the spring migration toward the traditional calving grounds.

Not all is peaceful as a herd of caribou travels and feeds on the hoof, because there is constant competition for the best feeding spots. In fact the conflict never stops. As snow becomes deeper or crusted, and the food underneath more scarce, the competition accelerates. Larger and more powerful animals drive smaller ones away from the few choice patches of food, by threatening and, less often, by kicking or a short,

Still in velvet, this caribou bull is nipping on willow tips as he travels over the Alaska Range foothills to the traditional rutting area. En route, he will gradually shed all of his antler velvet.

sharp antler clash. The loser gives up his feeding spot, probably to chase another caribou away from its place. Biologists have often seen caribou driving others from their beds as another way to establish dominance, which the winner can enforce later on in a feeding situation.

August, the beginning of the end of summer, finds caribou more widely scattered than at any other time. An observer seldom finds them in bands larger than a hundred. In Denali National Park, Peggy and I have watched bands of only a dozen or so when the land is aflame with autumn color. All of the animals seem to be calm since they are no longer pestered by the terrible insects of summer. They may stop to stand and rest often, as if in a torpor. They also seek out the lush vegetation that will soon wither with the frosts of fall. You might find caribou feeding around ponds and potholes at this time.

In Alaska and the Yukon, caribou often spend late summer feeding fairly high in the mountains. We noticed that in late August many of them gather in the high, barren passes along the North Canol Road on the Yukon-Northwest Territory border. Mornings at Old Squaw Lodge, Yukon, we would awaken to see many of them grazing the slopes all around our cabin, sometimes not far from a grizzly bear also feeding. Both the bruins and the caribou were busy acquiring the fat reserves they would so desperately need later on.

Fall Migration and the Rut

In early September, when there is a sharp bite in the air and perhaps a few snow flurries, the caribou bands begin another trip—the fall migra-

LEFT & BELOW
Often to a photographer, caribou seem to wander like dark ghosts across a landscape shrouded in cloud mists. I shot these photos in early autumn, in the foothills of the Alaska Range, when the weather, as here, is normally far from ideal.

As caribou travel, they constantly compete for choice feeding spots. Larger, more formidable animals dominate. The larger bulls dominate until they shed their antlers. After calving, the cows retain their modest antlers and use them to drive bulls, now antlerless, from favored food. Thus nature gives cows the critical edge over bulls when the cows must produce rich milk. (Karl Maslowski photo)

tion. This is subtle at first, but gradually gains momentum until it resembles a tide. The smaller bands begin to gather into herds. All make their way toward rutting areas, a trek that, in many cases, is southward toward the treeline. Elsewhere it might be a shift toward lower elevations. The caribou are looking sleek and fat, and no wonder.

Between July and late September, older bulls acquire 50 or more pounds of white fat in a 3-inch layer along their backs and rumps. This is an increase of 20 percent in body weight, which they will sorely need. A caribou cow's weight gain is more modest at about 10 percent of body weight by the beginning of the fall migration. But cows do not lose weight during the rut and may even continue to gain a little until November. All of the northern antlered and horned ungulates increase in body weight toward the onset of winter, but no species as much as the caribou.

Natives, hunters, and hunting outfitters have often told us that the first heavy snowfall is what triggers an acceleration of the fall migration toward its destination. I have witnessed only the first tentative gathering of the caribou bands and never the full migration, nor the rut. But in Denali, Peggy and I have watched the large bulls finally out of velvet begin the battles that will gain intensity as time passes. As the bulls' necks swell to almost twice their normal size, the bachelor friendship of summer turns into a growing intolerance of one another.

The furious contests of strength between males to do the breeding are the most dramatic incidents of the rut, if not of the entire caribou life cycle. According to George Calef, the dueling becomes more and more violent until, at the peak of the rut, the violence level is so great that some bulls are maimed or killed.

A rutting group is large, especially among the northernmost herds, and is composed of bands numbering to a hundred animals. The herd size is also very fluid, with bands merging, separating, and exchanging. The rut takes place on the move. Cows are not herded as with elk, but can move about freely. Males are constantly threatening and challenging, courting and searching. Trouble breaks out when more than one dominant bull begins to track the same cow in estrus.

A bull approaches an estrus cow with neck and nose extended forward while emitting a hoarse cough or panting sound, an attitude resembling a threat. The bull tries to approach from behind to smell the female's genitalia. Calef noted that as one bull thus follows a female, other bulls may quickly appear. If turning of the bull's head to display the massive size of his rack does not discourage these other suitors, a fight results. It is not unusual for multiple fights to break out. And Calef has found bulls with antlers locked, a situation that meant a slow and painful death.

The Rut's Toll on Bulls

Somehow despite the frenzy, confusion, and injuries of the rut, breeding is completed by the best of the bulls, and the cows are carrying embryos of the most determined and aggressive sires. But the great bulls must begin winter with fat reserves almost gone.

Bulls virtually stop eating for the duration of the rut. They develop a rank, stinking odor. (Native hunters know that old bulls are inedible during and after the rut.) Fat is used up. As glycogen stores are depleted, the liver degenerates, turning mushy and yellow. For a precarious time

Two young Peary's (barren-ground) caribou males duel in preparation for the years when they will be serious contenders for the herd's breeding privileges.

following the rut, the handsome, noble bulls of summer are reduced to skin, bruises, and bone. They are actually as vulnerable to predation as their own progeny, which were dropped just a few months before and a few hundred miles away. For the cow of the species, however, it is an entirely different story.

When Cows Become Dominant

Bulls are the dominant animals in the herd until they lose their antlers in late fall, after the rut. Pregnant cows retain their antlers, and suddenly they are the socially dominant caribou. They use their dominance in many ways, especially to obtain the high-nutrition foods necessary to nourish fetuses, which grow rapidly despite the bitterness of winter. The eventual separation of males from females into bachelor herds may well stem from the bulls' inability to compete with antlered cows. Thus evolution has helped the cows survive at the most critical time.

A female's antlers also come in handy later on for protecting calves, and for assuring them of places to feed. Calves inherit the social position of their mothers, and they are able to feed alongside them without having to dig on their own. If a cow dies or is killed, for example, the calf is shuffled toward the bottom of the herd's dominance order, where its chances for survival on its own are greatly diminished. When a cow no longer wants to nurse her own calf, she simply walks away from wherever she is feeding to allow the calf to have that spot.

Maybe the most important lesson a calf learns from an alert, protective mother is how to synchronize its life with the rest of the herd. Any animal that behaves differently—that wanders away from the herd or lags behind—is the one most likely to be preyed on by wolves.

Wolves, Man, and the Future

During winter and spring, Alaskan and Canadian wolves depend almost entirely on caribou to survive. No other sufficient source of food is then available. Thus the greatest danger to caribou during their spring migration toward calving grounds is the wolves that may migrate right along with them. George Calef believes that no other carnivore relies more completely on one species than wolves on the caribou of the northern tundra. Elsewhere to the south, as in Alaska, Yukon, and British Columbia, the wolves can also turn to sheep and moose, but in the barren lands these animals do not exist. During the spring migration or any other, the wolves kill those caribou that are weakest and least capable of maintaining the pace of the herd, simply because these stragglers are the easiest to catch. Wolves also capture a good many calves on the calving grounds. Figures compiled by several biologists reveal that one in every four calves will not survive one month during a normal year.

Of course, wolves are not the only killers. Native peoples of the North may have already replaced wolves as the major predators of many of the northern caribou herds. The meat is tasty, nutritious, and sometimes larded with fat. Caribou skins are so highly treasured throughout

Wolves and caribou actually depend on each other for survival. Wolves normally kill the young, aged, and weaker caribou, maintaining a strong and viable herd.

the Arctic for clothing and sleeping bags that in some native communities caribou bulls especially are killed more for the skins than for the meat.

Fortunately, the future of all caribou herds depends much more on the quantity and quality of calves produced than on how long the bulls live. I hope that all of the populations of caribou roaming North America today will always wander their remote and mysterious trails. Yet that may not be possible if people continue their crazy, indefatigable effort to conquer the wilderness.

Early in October 1984, an immense herd of eastern Canadian caribou were making the annual migration from Labrador toward Hudson Bay. The rut was also beginning. But at least 9,000 and perhaps as many as 22,000 did not reach their destination. The animals were drowned trying to cross the same stretches of treacherous rapids on the Koksoak and Caniapiscau rivers they had crossed every fall for centuries past. Why didn't they make it this time, too?

Eskimo leaders and Canadian conservationists blamed it on Hydro-Quebec, the province's government-owned, massive dam and hydro-electric facility on the Caniapiscau. They blamed the deadly river current on too much water being released, without warning, at a time of year when river levels are normally low—also the critical passage time of the caribou. For many years, northern Eskimos (who depend on the herd for subsistence) had asked the Quebec government to build a barrier to divert the herd from that dangerous crossing and better regulate water flow levels. They were ignored.

Belatedly the barrier had to be built in haste and at much greater cost. Aircraft also had to be used to help herd the caribou away from the worst fords to prevent an even greater tragedy. A large-scale removal and mass burial also had to be organized to prevent a serious pollution problem downstream. Concluded Canadian biologist Alexander Banfield, "At least the wolves will have enough to eat this winter."

It would be worthwhile if we really learned a lesson from that experience. But I wonder if we will.

4.
WHITETAIL DEER

It is a bitter December morning in south Texas. All night long a wild blue-norther has scoured the dry brush country, rattling the brittle yellow vegetation and driving dust. A powdering of snow covers the ground and the temperature is hovering at 25 degrees, although it feels much colder. This is a perfect morning for what my wife Peggy and I have in mind: ambushing a great whitetail buck that has been seen here lately.

Peggy and I park our pickup beside a concrete water tank and a defunct windmill that is barely visible against the predawn sky. Quietly we pack cameras, a thermos of coffee, groundcloths, and a secret weapon into our backpacks. We walk softly down a thin trail through mesquite and blackbrush.

The trail is familiar because we found this spot several days before. After studying the abundance of deer sign—fresh scrapes and huge hoofprints all around a small clearing— we had hastily built a blind on the spot. We would return later and try to photograph the deer that made the hoofprints. For now, we wonder if we will find the blind intact after the all-night storm.

We reach the blind, which is still standing, and Peggy crawls inside. While she sets up a camera with telephoto lens on a tripod, I find another spot about 40 feet downwind, where I can crouch, hidden in the center of clumps of prickly pear.

I have been photographing whitetail deer and other big game since the 1940s, yet the thrill remains. When I stalk wildlife, the old adrenaline pumps, I forget complaining muscles, and I never notice the cold as much as I do at other times. Now I sit, tense, listening to the sounds of a new day. Not far away a coyote yips. In the distance I hear the less

This gentle, timid fawn of 3½ months will soon enter the ranks of the adults of its species, known for extraordinary toughness and cunning.

PAGE 78
Many consider the mature whitetail buck the most handsome large mammal in America. This splendid, symmetrically antlered ten-pointer has the swollen neck of a buck in rut.

musical grunting of cattle, which are probably moving toward a water source. A Harris hawk glides past, probably looking for the covey of scaled quail that is calling nearby. In the brush just behind me, a dull-red cardinal-like bird called a pyrruloxia rattles about. Eventually the sun bathes the landscape in a bronze glow.

Checking with my exposure meter, I decide that there is minimum light for photography and set my camera dials accordingly. Then I take out my secret weapon—a pair of discarded whitetail deer antlers—and begin what I hope is a convincing performance.

I rattle and rub the antlers together, trying to imitate the sounds of rival bucks dueling. After a pause, I scrape the antlers on the ground, pause again, then rattle the antlers against brush. I touch the antlers together once more, then sit back to wait—but not for long. Because now I can hear a buck hurrying our way long before I can see him.

Suddenly the buck enters the clearing and stops between Peggy and me. The phenomenon is like the projection of a new slide onto a screen: One moment the scene is empty, a split-second later it contains a deer. I count six points on his head, and when he turns toward the sound of Peggy's camera motor drive, I raise my camera and focus on the sun's reflected highlight in his eye.

For several seconds the young buck stands uncertainly before us. He is puzzled and, in a deer's way, probably wonders where the sparring bucks are. I squeeze off several exposures before the buck becomes suspicious. He finally wanders away, no doubt disappointed. I am somewhat disappointed too, because this is not the buck that made the hoof-

Scenes like this of majestic whitetail bucks occur all too rarely for the deer watcher or photographer who simply wanders in the woods. For best results, erect a blind near a center of deer sign and activity and then wait.

prints all around us. With less hope than before, I scrape the antlers together again.

For 10 or 15 minutes nothing happens, except that the sunlight is much brighter and I adjust the exposure on my camera to compensate. I scrape the antlers on the ground and then crack one against the other. More time passes. By now my knees are cramping from being in a squat too long, so I shift my position ever so slightly. That is my great mistake.

For some time, the huge buck we are after has been standing not 50 feet behind our hiding places. He approached that close in total silence. I turn and catch sight of his neck, swollen with the rut, and his massive widespread antlers before he disappears. The pounding sound I hear is not of the buck running away but of my own pulse.

Like so many other mornings when I've pursued whitetails, that sudden brief encounter several years ago provided a lesson. It was also typical of a day in the life of a whitetail photographer. All creatures, from garter snakes to giraffes, are fascinating and at times very difficult to bring into sharp focus. But none are quite as challenging, nor quite as elusive, as the species of deer that is so familiar to all Americans.

Whitetail Origins

Fossil remains indicate that the swift and graceful whitetail, *Odocoileus virginianus*, evolved from a piglike ancestor, which millions of years ago stood less than 2 feet high. If you doubt it, consider the similarities between deer and swine today. Both have split, cloven hoofs, the prints from which someone inexperienced in the woods may be unable to tell apart. Both have elongated skulls and compact bodies, as well as short and stiff body hair. They share a keen sense of smell. Their legs are similar in structure, except that a deer's legs are longer. Deer and hogs also give multiple births. When a domestic hog becomes feral, it prefers the same habitat as whitetails. Both are favored prey by large predators.

Neither deer nor pigs sweat profusely when pursued or during any great physical exertion. Their vocalizations can be remarkably similar, and they both live on highly varied diets.

Adaptability

The whitetail deer is found wild in every state but Alaska, in addition to most of Mexico and southern Canada. Except perhaps for coyotes and raccoons, no large native American mammal is more adaptable than the whitetail to life in the late 20th century—in physiology and behavior as well as genetics.

Although whitetails first evolved into woodland browsers, which mainly they still are, they can adapt to eating anything from corn and alfalfa to soybeans and sweet potatoes. I have found well-nourished deer in such varied habitats as cedar bogs and canebreaks, mountain meadows and river floodplains, farm fencerows and even forests that have been scorched by wild fires.

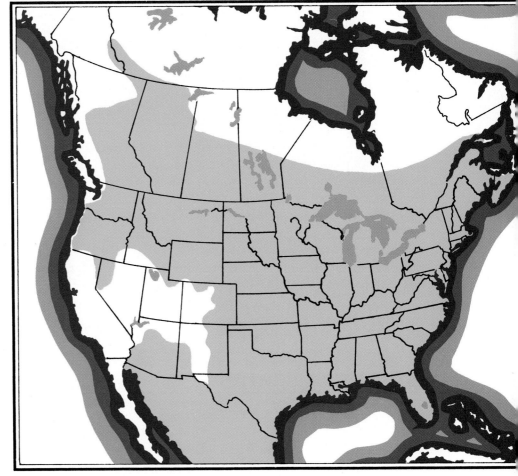

Whitetails seem to thrive best, however, along the irregular "edges"
or borders between different kinds of habitat, especially if there is heavy
cover nearby. In fact, whitetails have adapted so well to changing habitats and land use that they are far more numerous today than when Europeans first arrived and found a dark, unbroken forest covering much of
the land. That adaptability can be credited to an adaptable digestive
system and the whitetail's ability to cope with humans.

The Physical Specimen

A deer's brain is smaller relative to body size and much smoother than a
human brain, indicating that deer are not nearly as intelligent as people.
But sometimes that is difficult to believe.

Whitetails react instantly, and usually correctly, to sudden danger.
No matter how rapidly a deer is running, it can suddenly change directions without slowing down, a maneuver that holds enough shock to
break fragile front legs. But the rugged anatomy of the species make
this and other acrobatics routine. The whitetail's forelegs do not connect
directly to the rest of the skeleton. Instead they are separated by a
tough, resistent cartilage that serves as a shock absorber.

For me, it is impossible to spend time among whitetails and not

PAGE 85, TOP
The sharply tuned ears of a deer are constantly moving and can be independently aimed, or "focused," on sounds in all directions.

PAGE 85, BOTTOM
This bedded buck is a fine specimen of a high-antlered whitetail. The bed site is beneath the crest of a low hill so that the animal has a good field of vision all around. The buck was on his feet and heading away a moment after we shot the photo.

marvel at them. No other large native mammal can vanish from a scene so swiftly and so beautifully. Indeed, no other animal described in this book is equipped with as many efficient escape mechanisms. I figure I've been a victim of all of them.

A whitetail is fleet afoot and is a powerful leaper. True, the pronghorn antelope can generate greater speed in open country. But I doubt anything can match the speed of a really frightened whitetail plunging headlong over deadfalls and rockpiles, through a dense second growth forest—all in seemingly effortless bounds of 25 to 30 feet. Our files are full of photos of blurred whitetails running away.

The whitetail possesses excellent hearing, a phenomenal sense of smell, and good eyesight, even in semidarkness. Many times while walking toward a blind in a dense woods well before sunrise, or leaving after sunset, I have flushed deer and stopped to hear them crashing away, apparently unharmed, as if in broad daylight. The night vision of deer in many areas must be well developed, because the animals seldom move during the daytime.

A whitetail's vision may be slightly flawed, however. Go into a whitetail woods and sit or stand beside a busy deer trail. As long as the wind does not betray you, an approaching deer may come very close without spotting you—until you move, ever so slightly. Deer can instantly pick up movement, but a stationary object, even a strange one, can go unnoticed. Unfortunately, when photographing deer, it is impossible not to move head and hands to operate the camera.

Whenever you are out among whitetails, watch the ears in particular. The ears of even the calmest doe turn from facing forward to the sides and rear, keeping her constantly in touch with her surroundings. Without looking up, a whitetail can keep track of other deer in its own herd as well as detect the arrival of an intruder, usually from a great distance and with plenty of time to escape. Even though a deer may not show any visible sign, it is always alert and ready to flee.

We know from careful observations that a whitetail can distinguish between the sounds of another deer walking 50 yards away and, say, those of a bear or an elk. The same deer will also be alarmed or reassured by the movement of other deer nearby. A slow movement is mostly reassuring. But a fast movement will cause the hearer to look up in alarm. All of the normal woodland sounds to which a deer is attuned—many inaudible to humans—keep the animal informed.

Deer do not always choose to run, even when under great pressure to do so. I'm convinced that many of the oldest, most experienced animals rarely run at all; if possible, they just drift away out of sight before an intruder arrives. I've often seen deer stand motionless, or drop down into a crouch to allow people to pass nearby. I have often wondered exactly how many such motionless, camouflaged deer I have passed myself.

Early in the 1960s I decided to spend opening morning of the deer hunting season in the vicinity of a farm woodlot in Morrow County, Ohio. Since many hunters were afield that day and since I could hear shooting in the distance all around, I decided to play the waiting game. Being younger and much nimbler then, I climbed far up into an oak tree until I found a fairly comfortable place to sit. Almost immediately I

spotted the glint of eight-point antlers in a thicket bordering the stream that drained my woodlot. The exact time was 8:45 A.M.

During the rest of the morning, while I watched from my increasingly uncomfortable perch, four different hunters stillhunted within 50 yards of the deer, and one hunter passed within 5 yards! But the buck did not move an inch. He probably didn't blink an eye. I began to wonder if he was alive. Shortly after noon, I decided to try to stalk the buck myself, hunting carefully upwind in its direction. What I found was only a warm bed and moist pellets. Since then I've speculated if he was the same buck that was killed three autumns later by a beer truck on the highway not far away. This roadkill had the splendid trophy antlers that only a deer with great survival skills could develop.

When photographing a whitetail doe nursing a newborn fawn, or watching a herd of graceful, cautious deer grazing at dusk, it is easy to regard these animals as frail and defenseless. But those views are misleading. Even the best human athlete in prime condition couldn't match the strength or vitality of an average whitetail doe. It is impossible, for example, to catch and hold a healthy whitetail for tagging or scientific examination without the use of drugs or other catch-and-hold devices. Deer can also suffer severe wounds and even broken legs, yet somehow manage to survive.

A biologist friend once showed me a deer carcass imbedded with

Whitetails are constantly alert. Even when feeding busily, one or two does in a group will look up suddenly from time to time to check for suspicious movements or respond to sounds.

the broadhead and part of the shaft of an arrow, as well as the rifled slug fired from a 12-gauge shotgun. Both nearly lethal wounds had completely healed over, and at least a year after the last of the wounds, the medium-sized buck showed no sign whatever of the metal in its body.

In 1984 a rifle hunter shot a remarkable deer during Wisconsin's open season. On dressing the animal, he found the broadhead of a hunting arrow imbedded three-quarters of an inch into the brain. State biologists concluded that the deer had been shot as a small buck and lived 1½ years carrying the broadhead, which eventually became encapsulated in bone tissue.

Did those deer feel pain? Probably, though all wild animals have a much greater tolerance to pain than do humans. The acceptance of hardship—intense cold, lean times, hunger, even pain—is the way of life in the wilderness.

Feeding Habits

Whitetail deer seem to eat almost anything that grows where they live. They greatly prefer some plants to others, however, and will resort to certain least favorites only when everything else is gone. A deer's diet varies so much across America that listing deer foods here would be futile. But this much is certain: an average whitetail needs about 6 pounds of good quality browse a day to survive in optimum health. Of course the daily intake will be greater during the abundant days of summer, and much less during winter's hunger moon. Once as a test, Peggy and I picked for four hours to collect a typical day's deer browse. It amounted to about one full bushel basket.

Summer is a time of plenty for America's whitetails, and they take advantage of the bounty. Does recover the weight they lost during the winter, and if the range has not been overgrazed by domestic livestock, they grow sleek in their seasonal reddish coats. Fawns soon double and redouble what they weighed at birth. In a surprisingly short time fawns are able to follow their mothers almost anywhere. When summer ends they are self-sufficient, or nearly so. But the greatest change of all takes place in bucks.

How Antlers Grow

Toward the tag end of spring, a buck's pituitary gland—rather than male hormones (since a buck's testicles are inactive in spring)—causes antlers to begin growing from pedicels on the skull. The new antlers first appear in April or May, and grow from the tips outward. Covering them is a soft velvety membrane that is highly charged with blood vessels. These vessels carry chemicals that permit the antlers to grow, eventually to impressive size on some 5½ to 6½ year-old bucks. Any antler in velvet is tender, sensitive, and subject to injury. The velvet will bleed if cut. So even a large prime buck is a docile animal during the summer, avoiding damage to the antlers. Injury to an antler during this season

PAGES 88 & 89
As if getting within photographic range of a fine whitetail buck bedded down isn't rewarding enough, to our delight a second ten-pointer strolled into the scene.

results in a deformity that shows up on the antlers of following years too.

The antlers of healthy whitetail bucks reach maximum size in September when a new supply of testosterone in the blood "ripens" the antlers and stops the summer-long growth. When the blood supply is cut off, the velvet dries and is gradually discarded.

For a long time it seemed to me that bucks tried to rub off the velvet on brush and tree trunks perhaps because the antlers itched. But after watching many bucks more closely through long telephoto lenses, I now believe this shadowboxing is simply the first manifestation of the rut not too far in the future. We photographed one fine Ohio whitetail buck that fed in an apple orchard for several days. All the while, strands of velvet hung in front of his face like a bedraggled hairpiece, seeming to cut off his vision. But the buck was too busy eating to be bothered with rubbing.

A normal whitetail's antlers are symmetrical or nearly so, and the actual conformation or shape of growth is hereditary. A buck has the same basic antler conformation year after year. The only change is an increase in antler size as the animal reaches its prime at about 5½ or 6½ years old, and a decrease beyond that point.

Two main factors contribute to those splendid, heavy racks of widespread antlers that a few whitetails carry in autumn. Heredity is one. Another is the quality of the range on which the deer live. The range must provide a heavy diet of calcium, especially during the winter, as many researchers now believe. There seems to be an important link

Antler rubs on low trees and brush begin in early autumn when the velvet is discarded, continuing through the breeding season. During the rut, the buck will often make a ground scrape directly beneath the rub. He may also urinate in the scrape.

between the quality of soil and winter forage and the size of antlers attained the following fall.

ANTLER SHEDDING.   The antlers of any deer—or, for that matter, elk, moose, or caribou—are a solid bone structure shed and renewed every year as long as the animal lives. By contrast, the horns of wild sheep, goats, bison, and muskoxen are not shed, but continue to grow in size until the animal dies.

The Ohio buck mentioned earlier kept its antlers well past the rutting season, or as long as testosterone surged in his blood. I watched the animal rattle its bare antlers many times against saplings, and once, without fear or injury, to drive a slightly smaller buck away after a brief noisy clash. When the supply of hormone in the blood ebbs, the antlers fall off, usually in mid-winter. In winter, the former monarch of the region is reduced to a nonentity that must compete on a more equal footing with other deer to survive.

Trophy Bucks

Those large male whitetails commonly referred to as trophy or record-book bucks are not evenly distributed over whitetail range. They come from scattered and sometimes isolated areas where the genetic and calcium requirements are just right. Rarely do trophy bucks originate where there is a heavy population of deer, and never where the range is over-populated. Bucks with the largest antlers are currently found in the Midwest, particularly in Missouri, where two of the heaviest racks ever recorded were found.

For many years Peggy and I photographed whitetails on a south Texas brush country ranch, where conditions were virtually ideal for producing whitetail bucks with unusually large antlers. The owners accomplished this with sound management techniques, which included keeping the overall deer population well below the carrying capacity of the ranch, shooting all single-point spike bucks each fall because they are genetically inferior, and sparing the few super-buck breeders that roamed the area. Sitting in a blind on this ranch at daybreak one morning, I watched two 12-point bucks and one outstanding 14-point buck in view all at one time. Only one of the 12-pointers strolled into decent camera range, and then only after rain began to fall.

Fighting

Quite often, as I watched several deer feeding in a small area, one deer would suddenly kick out at the flank of the one standing nearest. Or more often, two deer, with ears dropped or flattened, would abruptly rear on hind legs and strike with forehoofs at the head of the other. Does as well as bucks would fight this way.

COMPETING FOR FOOD.   Competition for food, especially where it is scarce, is likely to spark conflict. Where northern deer are

Whitetails are not the submissive, Bambi-like animals they may seem to be. Does slash out with sharp hoofs at one another and even at fawns in disputes over food, social rank, or simple right-of-way on the trail.

concentrated in "yards"—particularly when they are fed artificially by man—there is bound to be trouble. Stronger deer drive smaller, weaker ones away. Does will not tolerate their own fawns when the going gets really grim and the temperature in February hits rock bottom.

Even when food is plentiful, whitetails fight for other reasons. Social rank in any herd must eventually be established; if bluffing or passive submission doesn't work, a physical confrontation is the answer. Biologists believe that the most confrontations occur when unfamiliar animals meet or are mixed, as might occur following the hectic period of a hunting season. In established herds, social rank is established early. All the animals "know" each other, and squabbles are few. But shortages of food and water can drive deer to areas already occupied by others. Biologist Edwin D. Michael, who made a two-year study of whitetails in San Patricio County, Texas, believes that a certain amount of aggression may be necessary to maintain a high reproductive rate.

On two occasions I have seen whitetails—Ohio animals with heavy racks in both cases—with antlers locked as a result of fighting. A team of game wardens were able to separate and save both bucks in one lockup, but they were able to save only one of the combatants in the other. In the less successful effort, the buck they saved charged them as soon as he was free! Probably most of the time both bucks in death locks die lingering deaths because the incident happens far from where anyone will discover it.

Minnesota forester Tom Swedlund once wrote to me that he had shot a large buck with what appeared from a distance to be a strange, abnormal rack. Approaching his kill, he found that it was carrying the rack and skull of another deer entangled in its own antlers. The winner had somehow outlived its rival and managed to wrench its head from the carcass. Swedlund's unique trophy was too thin and emaciated from the experience to be used as venison.

Antlers of these bucks locked because they were sprung open slightly in combat. In such cases, both bucks usually die unless one of them is able to exert enough force to unlock himself. The loser in initial combat usually suffers a broken neck; the "winner" may then fall prey to starvation or predators.

Dominance and Mating

Biologist Ed Michael has noted that bucks begin fighting well before the rut, at first by rearing and striking with hoofs, which he considers sometimes the most vicious, and later in head-to-head combat with antlers. Despite the sound and fury, serious injury seldom resulted, and this jousting may have gradually established dominance among males by the time the rut began. Michael also feels that early sparring and fighting tends to eliminate antler-locking incidents that may otherwise take place later.

The sharpest clashes I've witnessed were initiated by the presence of a doe in estrus. Any breeding bucks within scenting range of the doe are aggressive toward one another, each trying to keep between the doe

and his rivals. According to Michael, does have little or no choice in the process and breed with whatever buck is with them when they become fully receptive. Unfortunately for researchers, copulation usually takes place at night. After observing 130 instances of bucks actively pursuing does, Michael witnessed only one daytime copulation.

PAGE 94
During the breeding season, this buck was walking along a deer trail past my blind when it suddenly caught the lingering scent of a doe that had passed earlier.

Birth and Predation

Most whitetail females give birth to single spotted fawns. But where the habitat is good, or where the range has not been filled beyond its carrying capacity by deer, twin fawns are not uncommon. On two occasions, both in Ohio during the 1950s, I saw does in early autumn with triplets. But the odds are low for all three triplets to survive the critical first month or two of their lives, or even until they are virtually self-sufficient. There have also been records of whitetail does bearing four fawns at once, though never have all four survived.

The individual whitetail doe is so prolific that a whitetail herd can quickly grow to a size that overpopulates and damages its environment. (Biologists are puzzled about why well-nourished whitetail does produce more female offspring, while undernourished ones drop more males.) In theory, one whitetail buck and one doe in a predator-free habitat could become a herd of 22 deer in five years. Projected further, theory says this same herd could number 189 in just a decade. With these figures in mind, it is easier to understand the argument for hunting seasons in almost all of the United States. In Michigan alone, it has been estimated that about 500 cougars or about 750 wolves would be required to harvest the same number of deer as hunters do in order to keep the herd in healthy balance. It is unfortunate that cougars and wolves have been eliminated from so much deer range throughout North America, though in most places their return is out of the question.

Doe and Fawn

Right after birth a whitetail fawn is thoroughly licked and nuzzled by its mother. The mother also eats the placental sac. For a day or so she usually remains with or near the newborn. After that the doe spends only brief periods every day with the fawn, which lies motionless in the woods while she feeds. The mother returns at intervals to nurse the fawn. It is during this period when fawn mortality is highest. Such predators as bobcats, coyotes, and bears, as well as humans and their free-running dogs find too many helpless fawns. Some people think that because it is alone a fawn has been abandoned. Few wild young creatures are ever abandoned.

In time the wild mother teaches her fawn by example when to run and when to hide, how to nibble on which kinds of nutritious woody plants, and how to constantly test the wind currents all around for signs of trouble. Older, experienced mothers are likely to be the best teachers because their own survival skills have been effective longer. Just know-

The survival rate of whitetail fawns is not high in any part of North America. They are most vulnerable soon after birth. This fawn is 3½ months old, which means the odds for its survival have greatly improved.

ing when to stand motionless or to flush in an instant can be the difference between life and death.

Recently, while hiking and photographing woodland plants in a northern Minnesota woods in spring, I stooped down for a close look at yellow moccasin flowers. But just a few feet away I spotted something far more interesting, something I could not see when standing up. A whitetail fawn, possibly less than a day old, lay absolutely motionless in green vegetation.

The fawn did not even blink an eye or show any visible signs of stress while I shot a few closeups before I hurried away. Try to imagine similar self-discipline in a human baby, or in a child of any age.

Fawn Mortality

According to recent field studies completed in Missouri, a fawn is more likely to perish during the first month or two of its life than a fully mature buck is during an open hunting season. A fawn that survives its first year is thus already a very wary animal. As it grows older, it becomes warier, especially if it lives in a region where hunting takes place every fall.

Any fawn's survival and later behavior depends on its mother. If she is a wild, alert and suspicious animal, the fawn will also be wary and shy. By contrast, a doe that has been raised within frequent sight and sound of humans, or even in outright captivity, will have fawns that are less wary or even quite tame. Experiences later in life may alter a fawn's behavior, but the pattern is set early. For example, a fawn captured soon after birth and raised as a pet never learns the fine art of escape. But wild ones certainly do.

Domestication of Fawns

In most states, picking up a wild fawn is illegal. Nothing but disappointment, disease, and injury is in store for captor and captive when trying to domesticate a deer. Deer can carry such deadly parasites as virus-bearing ticks and fleas, rabies, tetanus, and tularemia. Adults, but mostly children, have been bitten, stomped, and kicked, and a few people are even killed every year by "harmless" whitetails "rescued" from their natural homes.

Endurance in the Chase

Especially in Canadian wilderness areas where gray wolves still hunt, these wolves prey on whitetails as they have for millenia. However, a good deer population always survives, and one way it does so is by swimming. In areas with many lakes and rivers, taking to the water has become an effective escape mechanism.

In November 1982, Michael Nelson was flying over a lake-studded portion of the Superior National Forest, Minnesota, following a pack of seven wolves. One of the wolves had been live-trapped and was radio-collared to study its movements. At exactly 10:45 A.M. Nelson noted that the wolves were running along a partially frozen lake on a course parallel to a swimming whitetail deer, apparently in an effort to intercept it. The temperature that morning was several degrees below freezing. About 100 yards from shore, the deer seemed to realize that it had been

Wolves have evolved with, and are important predators of, many large native mammals. They are also important citizens of the wilderness. The wolves in this photo are male siblings.

Now after the rut, this big buck's antlers will soon be shed, leaving him looking much like a doe from a distance. The shedding of antlers itself, however, leaves a buck only slightly more vulnerable to predators because a whitetail's first defense is high-speed flight. The main hazard to a buck after the rut is his depleted fat and energy reserves, which may not sustain him through a hard winter.

cut off and it turned to swim back across Thomas Lake, a distance of over two miles.

Nelson had to leave the area to refuel, but he returned at 1:15 P.M.—3½ hours later—to find the whitetail still swimming and the wolves still stalking from the nearest point on shore. One wolf entered the water, swam, and drove the deer to the shallows surrounding a small island where the deer rested briefly in the shallows. But then the deer was forced to swim for another hour, eventually becoming weak and barely able to move. One of several wolves swimming in pursuit killed the deer in the water and dragged it to shore.

Consider the deer's extraordinary tenacity: It had been swimming at least 4½ hours after the pilot first spotted it, in nearly freezing water. Of all land mammals, only polar bears could be expected to survive such terrible exposure.

The Wolf as Prey

Occasionally whitetails also kill wolves. Biologists have found numerous wolf skulls with injuries inflicted from severe blows. Most were struck by moose, but not all. Some of the blows almost surely came from deer hoofs. In December 1983, near Silver Island Lake in the Superior National Forest, Tom Pearson found a wolf carcass atop freshly fallen snow. About 40 yards away was a freshly killed nine-point whitetail buck. Pearson was able to reconstruct exactly what had happened from tracks and signs in the snow.

Two wolves had chased the buck down a forest road while two others broke out of the woods to intercept him. Signs of the four-on-one attack marked the spot. Not far away was a wolf bed with blood on it. Tracks from the bed led to the place where the carcass had been found. Blood on the buck's antler tines revealed that he had gored the wolf to death before being overwhelmed.

The dead wolf was a female of about 75 pounds, in sound health with good fat reserves. Although most edible parts of the deer carcass had been consumed, it had no skeletal injuries, and fat content in the bone marrow was high, an indication of the buck's good health. The buck was not an old or weak animal especially vulnerable to wolves; he simply couldn't handle four wolves. The deep, fresh, snowfall may also have hindered the buck's ability to escape by running, which is its major escape tactic.

Whitetail Longevity

With or without hardship, deer live fairly short lives. In captivity, the very oldest animals may reach 15 years, but purely wild whitetails rarely reach that age. In states with larger deer populations and regular hunting seasons, deer seldom survive beyond their fourth or fifth autumn. Dis-

Older whitetail bucks do not normally associate with groups of does and fawns. But this is the breeding season, and one of the females in this small band is probably coming into estrus.

eases, prolonged severe weather, predation (especially in winter and from domestic dogs), poaching and highway kills, as well as lawful hunting, assure a short life cycle in the deer population.

Although large antlers on a buck can indicate that the animal is older than a buck with smaller antlers, that is not always the case. Nor is the number of tines a reliable index of age. The best indication is found in the animal's mouth. An experienced biologist can fairly accurately determine a deer's age from the progressive wear and condition of its teeth. Exact age can be determined by examination of annual tooth rings under a microscope.

Too Many Whitetails?

Practically in the shadow of Chicago around busy O'Hare International Airport, a whitetail herd has grown to 100 per square mile. Although the roadkill in that county alone is 400 whitetails a year, even that is not enough to keep the population from exploding. An estimated 6,000 animals live on 65,000 acres, all of it closed to hunters. Live netting and removal procedures have not been able to compensate for the annual population increase, and the problem has reached crisis level.

Of course too many deer is not the problem everywhere. Many Americans wish they could see more whitetails around their homes and where they hunt. Larry Weishuhn, a busy and energetic Texas deer biologist, maintains that both the number and quality of whitetails can be increased and improved by planting small food plots especially for deer.

Planting a Food Plot

According to Weishuhn, sowing numerous small plots is preferable to one large one. Mixtures of oats and clover planted in the fall are good. Wheat is better nutritionally, but in many areas whitetails still prefer oats or alfalfa. Those of you living in rural areas should have soil samples analyzed for nitrogen and phosphorus. You may have to fertilize if the soil isn't rich enough for planting.

In planning deer plots, remember that long and narrow plots with some trees and brush left intact provide cover, a feeling of security, and resting places. But do not kill off good naturally growing deer foods just to plant a food plot. Remember also to fence off the deer plots from livestock which can't clear a 4-foot fence. Deer can though, with 3 or 4 feet to spare.

The Future

The North American whitetail is a tough, adaptable, prolific species, and the smallest member of the deer family on this continent. Yet it has the brightest future, and probably the most admirers, of all the other hoofed animals in this book.

5.
MULE DEER & BLACKTAILS

The mule deer, *Odocoileus hemionus*, is a native of the western United States. I have hiked widely over much of the mule deer's range in search of the animal, at first to hunt and since the mid-1960s only to observe and photograph. The many miles on foot and on horseback have taken me to some of the most beautiful places in the Rocky Mountains, from New Mexico to Alberta. Except possibly for Alaska, there are no more glorious landscapes than in the national parks and wilderness areas of the Rockies. Yet, after all this vagabonding, I saw the biggest mule buck of my life within walking distance of my home.

In the south sector of Grand Teton National Park, a hiking trail runs from the White Grass Ranger Station to Phelps Lake. The trail normally accommodates hundreds of hikers every day throughout the summer. But it was deserted on the bitter, gray November afternoon when Peggy and I departed from White Grass and started up the steep, rocky path. We were taking one of our daily hikes to keep us in good physical condition. We carried cameras, as always, because there were often moose in the area at this time of the year.

We were barely underway when Peggy suddenly stopped and pointed to fresh deer tracks crossing the trail just ahead of us. The hoofprints were large and deep, indicating a heavy animal. I paused and looked around. On the slope about 75 yards to the right and above us stood a buck watching its backtrail.

For several moments we could not believe what we saw. The deer was massive in body, with swollen neck, but its rack was even more impressive. I have seen mature male elk whose antlers didn't match these. The main beams appeared as thick as my forearm and widespread, and they were extremely high. I fumbled for the camera in my pack, but I was much too slow.

These twin fawns are puzzled at the sight of me, beneath a backpack full of camping gear plodding along a forest trail. They stopped and stared just long enough for this photograph.

PAGE 102
I photographed this magnificent mule deer buck in the Canadian Rockies during the fall rut when bucks are most active and visible. He was in the prime of life. One of the photos in the series made the cover of *Outdoor Life*.

By the time I snapped the shutter, the buck was already far away. We do have one imperfect slide to prove that we were not hallucinating. Not only was it the largest mule deer I had ever seen in three decades of searching, it will never be surpassed.

We hiked that same trail every day for the next week, always stopping long enough to scan the surrounding snow-covered mountainsides. But we never saw that buck, nor even its tracks, again. Nor did we ever hear of anyone bagging such a trophy in the open-hunting area that surrounds Grand Teton Park.

We saw very few mule deer in that area until a year later. Again in November, Peggy and I were hiking nearby—this time on Granite Creek, along a trail that winds and switchbacks up into Granite Canyon. Again, not far from our starting point a doe bounced across our path. We stopped to watch, and a buck appeared just behind her. This buck also had a massive rack. Yet again, we managed to squeeze off a couple of indistinct exposures as evidence that the buck had antlers of extraordinary proportions, though of different symmetry than the first.

I now believe that was the third largest mule buck I have ever found. The second largest was an Alberta buck that made the cover of the August 1985 issue of *Outdoor Life* magazine.

Finding those two huge Wyoming bucks only a mile or so apart tends to confirm an old supposition—that the biggest mule deer, at least those with the finest antlers, are not scattered equally wherever the species exists. Instead they are found in just a few, sometimes small, areas where all conditions for health and growth are right. Almost without exception these are not areas where other mule deer are very plentiful.

PAGE 104
The body of this truly splendid Rocky Mountain buck mule deer is insulated with a layer of fat from high summer living. The rack also is heavy and impressive. Most of that fat cushion will be used up during the rut and the bitter winter to follow, however, while the antlers will fall away in December or January at the latest.

The shape, symmetry, and mass of antlers varies immensely among mule bucks, from high and narrow (which is rare) to wide with heavy tines. Mule deer racks average wider and heavier than those of most whitetails.

Here are ranges of the mule
deer and its distinctive subspecies of the West Coast the
Columbian blacktail and—to
the north along the Coast—
the Sitka blacktail.

Mule deer thrive and grow
large in mountain environments such as this in northwestern Wyoming. They
share the range, as they have
for centuries, with elk and
bears. This is scenic territory.

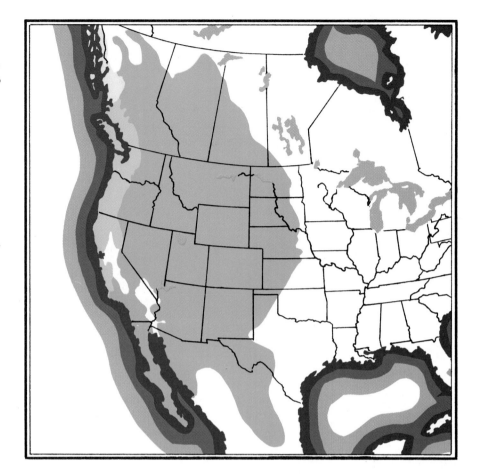

Mule Deer and Blacktail Origins

Originally mule deer lived in that area of the United States, Canada, and northern Mexico west of the 100-degree meridian and east of the Cascade Mountains. The Pacific coastal strip west of the Cascades was occupied by the closely related blacktail deer. Blacktails are usually regarded as a subspecies of mule deer, but just the opposite is nearer the truth. Both, along with the whitetail, can trace their lineage back to the same pig-like Pleistocene ancestor.

The true population of the mule deer when Europeans first arrived in North America is not known, but many modern biologists doubt naturalist Ernest Thompson Seton's estimate of 10 million. In fact, the number may have been only half of that. The logs of most of the early explorers and trappers mention few mule deer. One explanation may be that muleys were relatively unimportant to these people. True, the men needed wild meat to survive, but that was easier to get from larger animals, especially bison.

Muley Importance to Indians

Mule deer were important, however, to both the culture and subsistence of many western Indian tribes. Hardened antlers became the tool with which to work flint. Rawhides were converted by squaws into such soft, attractive buckskin products as moccasins and dresses, leggings and capes.

Although not nearly as desirable as buffalo fat, deer tallow was never wasted. It was eaten raw, used to process hides and to wax bowstrings. Bones were fashioned into needles and many other useful

I associate mule deer with trout and good fishing because the species is most abundant in wilderness areas where the waters are still cold, clear, and unpolluted.

tools. Many yellowed old accounts mention that when thinly scraped and stretched, buckskin could even be used as window panes.

Mule Deer Numbers

Although currently on a slow and steady downward population trend, mule deer are probably more abundant today than when the first wagon trains began to navigate the western plains. In a good mule deer state such as Wyoming, for example, about 421,000 mule deer roam on 49 million acres of suitable habitat—an average of one deer per 16 acres statewide. About 60,000 animals are harvested annually by twice that many hunters, after each hunter has spent an average of three days in the field. Density and ease of hunting vary, of course, from area to area. The trophy bucks we found near home in Jackson Hole—as well as the Alberta animal—were of the Rocky Mountain race of mule deer, by far the most widespread of eleven subspecies.

Range Shared with Whitetails

With tail flagged up in alarm, this whitetail buck is about to break away in a run characterized by darting and long bounds that will appear to make him float on air.

The ranges of mule deer and whitetails are overlapping more and more as whitetails seem to be expanding their range westward, often by following the river bottoms upstream and then establishing themselves around irrigated croplands. It is often possible to find both species in the same immediate area, if not in the same fields. They do not interbreed. After some observation, it is possible to tell them apart at a glance.

Distinguishing Between Mule Deer and Whitetails

The tails are distinctive. Whitetails have pennant-shape tails, 10 to 12 inches long in adults, brown above and white underneath. When whitetails are alarmed or running, the "flag" is often raised and waved, clearly showing the namesake white. Mule deer have shorter ropelike tails by contrast, whitish with black tips, which are seldom raised when the deer is frightened or running. Blacktail deer have dark brown or black tails.

EARS AND ANTLERS.   Mule deer have much larger ears (for which they are named) than whitetails, most evident when the two are together. The antler conformation of mule deer bucks is also distinctive. While a typical whitetail rack consists of opposing main beams from which extra points grow upward, the main beams on mule and blacktail deer branch into two beams, each of which branches yet again. Mule deer racks also tend to be heavier and more massive.

Mule bucks as well as any North American deer might, with some combination of age, diet, and injury, grow what are called nontypical antlers. These may grow out in strange shapes or directions, massively, often with many more points, twists, and drop points than normal antlers. Bucks with some nontypical racks are extremely impressive and sought after by trophy hunters.

WEIGHT.   Mule deer average heavier in body weight than whitetails. But some races of whitetails in some areas, especially in the northern

With tail characteristically dangling, mule deer run with a bouncing stiff-legged gait that seems awkward, but that is deceptive. Mule deer can cover rough ground with considerable speed, and in my mind a mule deer is at least as good a climber as any of the native wild sheep.

part of their range, may average heavier than some mule deer subspecies. At a research station in eastern Wyoming, mule deer bucks examined between 1956 and 1971 had average weights as follows: yearlings, 129 pounds; two-year-olds, 170 pounds; ages three to five, 206 pounds; six years and over, 249 pounds.

OTHER DIFFERENCES.   The metatarsal gland of mule deer is located on the hind leg halfway between the hock and hoof, and is covered with brown hair. A whitetail's metatarsal is smaller and is covered with white hair. But these can be observed only with difficulty.

Most of the time I can quickly tell a mule deer from a whitetail the instant it begins to move away—or run. The whitetail covers ground with a fluid lope similar to, but much more graceful than, most other hoofed animals. Mule deer bound away in a unique, stiff-legged gait that has been compared to that of a pogo-stick bounce. When genuinely alarmed, a mule deer will seem to bounce away by striking downward and backward with all four feet at once. That may seem awkward, but it is deceptively fast because a mule deer can easily cover 20 feet and more in a single bound. Far too often Peggy and I have heard mule deer escaping at our approach well before we saw them. Sometimes in heavy cover we never see them at all.

Diet and Change of Season

Winter is a tough time for mule deer. Most of their high altitude range is subject to savage storms, deep snowfall, and intense cold. In fact, winter weather can be a far greater factor in controlling deer numbers than the hunter's harvest. Snow depth covers many favored foods, so mule deer must turn to trees and shrubs for browse and for most of their winter diet. Not only is this browse relatively low in nutrition, mule deer must often compete for it with larger, taller elk and moose. The result is that most mule deer emerge from winter thin and in poor physical condition.

Spring is a reprieve. More and more high-nutrition forbs and succulent new shrubs appear. Deer nibble on more and more plants including the wild hollyhocks just sprouting in my backyard. A sleek new reddish coat appears, and fat begins to cover the ribs.

Except for horseflies, ticks, mosquitos, and other pests, summer is a halcyon time for mule deer, which consume large amounts of food for their size. Toward the end of the season, adult deer reach their heaviest weight. The red mule deer of high summer are beautiful creatures to see. In autumn, their red coat turns to gray.

When to Look for Mule Deer

More even than whitetails, mule deer are most active during early mornings and from late afternoons until dark. Daytimes are spent in cover, bedded down. Bedding areas offer concealment, an open view of approaches all around, and an escape route, usually behind and uphill. Anyone searching for mule deer for any reason must be willing to rise

Except during the autumn rut, mule deer bucks spend a good part of their lives bedded down. They look for sites well hidden or camouflaged, in the shade and at high elevation.

early and stay out late. The only exception to that might be during the late autumn rut when deer may continue their movement throughout the day. Mule deer may also remain active throughout the night during the breeding season. While camped in traditional rutting areas, we have often heard the restless animals all night long around our tent.

Mating

The annual mule deer rut, a rite of blustery, raw November in the northern Rockies, is a dramatic time we try not to miss. Does begin to funnel downward from the green pastures of summer to lower levels where they are more concentrated and more accessible. As days grow shorter and colder, estrus is triggered in first one doe and then another until it seems contagious. Bucks are soon attracted to the magnet.

For from two weeks to a month, bucks engage in a continual tournament to determine which hyperactive animal will prevail and breed as

This wandering, free-lance mule deer buck is beside himself with passion. As the buck travels through the woods during the rutting season, he pauses to slash at low bushes and sniff deeply for information whenever he crosses trails of other deer.

many of the does as possible. Sharp, savage antler clashes can occur when one rival moves in too near. In Yellowstone Park, we saw a medium-size buck blinded with a sharp tine in the eye from the sudden rush of a large, annoyed male. In this manner the best genetic traits are passed along from generation to generation.

We often see calendar paintings of a majestic mule buck standing amid a harem of females against a multicolored autumn background. But such depictions are false. Unlike elk bulls, mule deer bucks do not gather harems. In fact, you are far more likely to find several bucks tracking a single doe, the dominant or more powerful buck always keeping himself between the female and his rivals.

In many states the mule deer hunting season opens in October, just weeks before the rutting season. As a result, some of the larger bucks are killed before they have a chance to breed, and there is natural con-

cern that not enough males will be on hand to impregnate all the does. But there are probably few so-called barren does following a rutting season. A sex ratio as low as one buck per 12 or 13 does is enough to ensure adequate breeding. In a semi-captive herd under close observation in Alberta, one buck was known to breed 17 does with a resulting crop of 28 fawns. Scent glands located on the hind legs at hock and foot are activated by the rut, and these make it easy for animals ready to breed in locating one another.

There is also a compensating mechanism that nearly guarantees successful breeding: both males and females have the strong urge to re-produce. Even though the enthusiasm of dominant bucks may wane toward the end of the rut, biologists have noticed that the fervor of any unmated does will increase. Late in the season, it is not unusual to see an occasional doe aggressively following a buck.

Two small bucks spar head-to-head, probably to settle the question of rank between them. Neither is old enough or large enough to participate in the current year's rut.

Especially during the rut, I have found mule deer—in this case a breeding pair—high in isolated pockets on sheer mountain slopes.

A mule deer doe is attractive to bucks for only two days at most. Then her popularity wanes, and you must look to the next doe coming into estrus to find the bucks. One benefit of photographing during the rutting season is that most males lose some of their concern for man. A few become so enraptured that they all but ignore the busy photographer.

Mule Deer Births

Female mule deer become sexually mature during their second autumn, and an estimated 95 percent of them become pregnant each successive fall as long as they live. During the first five months after the rut, the fetus develops very slowly, but growth is fairly rapid after that. This sudden rapid growth usually coincides with the springtime crop of available browse. Does gradually separate to wander away from their herds for about two weeks to have their fawns in a secluded place.

Although we once found twin fawns on May 18, most Wyoming fawns are born during the first two weeks in June. A single fawn is usually the result of a mule doe's first pregnancy. After that twins are most common. Yet we have seen a few does followed by triplet fawns.

The Mother Instinct

At least some does are determined protectors of their young. D. W. "Ace" Wilcox, an old friend and veteran cowhand in western Wyoming, described a revealing experience to us. Riding fences early one morning, Wilcox saw a doe behaving strangely on the edge of a quaking aspen thicket. He dismounted, and through binoculars he saw that the commotion was the result of her trying to nurse a pair of nervous fawns. One of the fawns began to walk away on still unsteady legs and, while Wilcox watched, a coyote appeared and snatched the wayward fawn by the throat.

In the same instant the doe swung about and seemed to pounce directly on top of the coyote. At last the coyote dropped the fawn and ran into tall sagebrush with the mule deer right on top of it. Wilcox had the distinct impression from the way the coyote tried to escape that it had been severely injured on the first contact. The coyote may also have been young and inexperienced. Wilcox believes that the fawn was not seriously injured, because he saw the three deer together several days later.

Toward the end of the summer, fawns become increasingly self-reliant. The doe's protective instincts begin to fade away, although her fawns may keep on following her until the next spring.

Fawn Mortality

A number of times we have seen does chasing coyotes, probably in defense of young though we could not be certain. Despite any such maternal efforts, however, fawn mortality is high during the summer. Bobcats, wolverines, black bears, even golden eagles are among the

This coyote pauses in an area used heavily by larger animals. Notice how the trunks of the aspens have been eaten as high as elk can reach. This happens to be a deer and elk wintering area, and aspen bark is their last-resort survival food.

The cougar—also known as the mountain lion, or puma, or catamount—is the largest wild cat in North America. Where mule deer range, the cougar is a significant predator but will not wipe out a herd. When allowed to exist without bounties on their hides, cougars strike a natural, healthy population balance with that of mule deer and other prey species.

possible predators. But experienced biologists know that, except for isolated instances, normal predation does not greatly affect the health and prosperity of a mule deer herd. Habitat, with emphasis on winter habitat, or range, is what makes all the difference.

The Secluded Male

To see the mule deer bucks, especially the oldest and presumably wisest ones, you would have to head for the high country—away from the summer range of does and fawns, and also from the tracks beaten by human beings. A large mule deer is one of the hardest of all big game to find until summer is a happy memory. Mule bucks seem to spend the months with longest daylight hours in remote isolation in high country normally associated with wild sheep and mountain goats. More than once in high passes and alpine basins well above timberline, Peggy and I were surprised to find a great old buck, or even a small bachelor group of them, bedded down where bighorn rams should have been.

Not only do these bucks lead indolent lives on top of the world, gaining strength for the rutting season just ahead, they are agile enough to live in sheep and goat habitat. While backpacking one afternoon, we came suddenly upon a buck bedded beside an edge where a winter avalanche had fallen away from the side of the mountain. The area all around was a steep jumble of large, loose boulders and crisscrossed deadfalls. But that buck bounded from its bed, away from us and across that nightmare terrain as easily as over a grassy meadow.

In fact, Ken Clark of Fairview, Wyoming, is a mule deer hunting guide with almost phenomenal success season after season. He admits that the secret to his success is simply that he hunts in difficult, distant places, along thin ridges and within dark evergreens where other guides and hunters are unwilling to go.

After the Rut

The large breeding bucks do not join the wintering mule deer herds when the rutting season ends. Instead they resume life as bachelors, or join bachelor "clubs," with no interest whatever in family life. But the bachelor life of winter isn't exactly idyllic. The turmoil and constant competition of breeding leaves many of the most powerful bucks fatigued and in poor, if not emaciated, condition. They are easier victims of predators than at any other season. Mountain lions especially might kill a good many of them.

Road Carnage

Mule deer suffer from more than predation during the winter. In southwestern Wyoming for example, a section of U.S. 30 highway connects the Granger interchange on busy I-80 and Sage Junction near Cokeville. Every day in winter, as many as 1,500 large truck rigs use this 15-mile stretch of road through Nugget Canyon—through the same canyon that is a mule deer migration route from summer to winter ranges and back again.

During a rare mild winter, only about 300 deer are killed in Nugget Canyon. In an average winter the toll is 500 to 600 muleys. During severe winters the slaughter has actually exceeded 1,200. In 1984, more canyon-area deer were killed by trucks than were harvested by hunters.

Twin mule deer fawns get along very well until both want the same forb or morsel of food growing under the snow; then they do not hesitate to kick at one another.

An increasing number of elk, a few moose, and antelope also die in Nugget Canyon. Golden and bald eagles, too heavy to fly after gorging on the deer carcasses, also are unable to escape the traffic. The Wyoming Wildlife Federation recently convened a workshop of concerned agencies and biologists to seek a solution to the carnage by fencing, overpassing, or other reasonable means. It is hoped that their efforts succeed, but unfortunately Nugget Canyon is not an isolated situation in the Rocky Mountain West.

Emergence of the Blacktail

During their historic voyage down the Columbia River to the Pacific Ocean in the early 1800s, explorers Lewis and Clark passed through what today is good blacktail deer country. But the big-game animals they found in the Columbia bottomlands were elk. Clark noted that deer, themselves, were very scarce.

A blacktail doe is shown in a vast field of mountain lupine in Washington's Olympic Range. The Olympic National Park is an especially good place to see this species, which is more furtive elsewhere.

The mostly black, small tail of the blacktail deer distinguishes it from the mule deer, which has a dangling rope-like light-colored tail with a black tip.

Since then the wildlife picture in the Pacific Northwest has vastly changed. Elk are now harder to find and other native species have been eliminated. Generations of clearing, deforestation, plowing, and fence building changed the region so drastically that Lewis and Clark would never recognize it. The good news is that all this development actually benefited the Columbian blacktail. There are rich forage areas today in western Washington and Oregon with as many as 150 blacktails deer per square mile. This mule deer subspecies, *Odocoileus hemionus columbianus*, is one of those that have adapted best to civilization and cultivation. By contrast, modern wildlife researchers believe that the forest supported less than 10 blacktails per square mile in the late 1800s.

Blacktail Numbers and Range

There are about 2 million Columbian blacktails within their range—a narrow band that runs from the Pacific Coast east to the Cascade Range divide, and from northern California to British Columbia, depending on the season. The population is smallest at winter's end, while in midsummer it might be as high as 3 million. But despite their impressive numbers and concentration, we know less about blacktails than about other North American deer. Their shy secretive nature and ability to blend silently into their lush, often damp environment make them more difficult to observe and therefore study. That also explains why Peggy and I have fewer good photographs of blacktail deer in our files than of all other native big game.

Owing to their nature and terrain, blacktails can be hard to spot. At first glance, they can easily be mistaken for their bigger cousin, the mule deer, or even a whitetail. A blacktail's antlers are bifurcated (double Y's on each side) like those of the mule deer. On bucks of comparable age and body size, the antlers of the blacktail almost never match the size or mass of those of the mule deer. Instead of the rope-shaped, light-colored, black-tipped tail of the mule deer, the blacktail's tail is somewhat pennant-shaped, more like that of the whitetail, though smaller than a whitetail's. Also, the tail is dark or black on top, becoming darker toward the tip.

Feeding Habits and Survival Skills

Like the other large mammals described in this book, blacktails capitalize on abundant and nutritious forage for growth and weight gain from early spring through fall. In winter, when forage supplies are low and energy costs high, they save their strength by moving only as much as necessary. Fawns are born to coincide with increasing food supplies of spring, as are the young of most ungulates.

According to animal behaviorist Valerius Geist of Calgary University, Alberta, blacktails are masters at hiding and camouflage, taking advantage of whatever cover exists to remain unseen. When that doesn't work, or when a person wanders too close, the deer will seem to detonate as suddenly as a covey of wild quail, then bounce away on stiff legs. The peculiar gait, sometimes known as stotting, is very noisy in a leaf-littered woods.

Mammal behaviorist Dietland Muller-Schwarze believes that Columbian blacktails are unusually good learners compared to other ungulates, though not when compared to their predators. Their slightly larger brain relative to other hoofed mammals may permit them to cope better with emergencies and become more tolerant of the sight and sound (but not the nearness) of humans. Blacktails normally have more and slightly heavier evergreen cover in which to live than whitetails and particularly more than mule deer.

Blacktail Mating and Births

It is fortunate that blacktails are prolific animals. Biologists have found that only five or six bucks are needed in a herd of one hundred to assure that nearly all of the does will have fawns the following spring. What Peggy and I have witnessed of the blacktail rut in Olympic National Park is quite similar to the rutting ritual of mule deer. The more powerful or dominant bucks do most of the breeding, with as many does as time and traveling distance permit. The big ones do not try to gather harems and lesser bucks are able to do some of the breeding. The blacktail rut lasts from October until early December, running later in lower, warmer ranges. Fawns are born, after seven months' gestation, from May until the last ones drop in July.

We have never come upon a new blacktail fawn in the wild, though we have come upon whitetail and mule deer fawns. That may be because we have not spent as much time in blacktail country, and also because the species may be more reclusive. While hiking a trail in Washington's Olympic National Park one June morning, however, we spotted a doe and fawn in a lush meadow below us. A moment later the doe saw us and immediately walked away from the spot, not followed by the fawn. We then made a hasty search for the youngster but failed to find it.

Like the other fawns, blacktails spend their first weeks of their lives hiding, waiting camouflaged and motionless while the mother forages. Somehow the white-spotted reddish coat is hard to detect even in the new green vegetation of spring. Because predators will discover a fawn by scent rather than sight, a doe immediately eats all the vegetation, as well as the placenta, around the birth site with the result that no odor is left behind. Despite these efforts, up to 70 percent of newborn fawns might be lost in a normal year.

The Sitka Blacktail

Another race of blacktail virtually indistinguishable from the Columbian blacktail of Washington and Oregon is *Odocoileus hemionus sitkensis*, the Alaskan Sitka deer. This one ranges from the Queen Charlotte Islands, British Columbia, northward through southeastern Alaska to about Yakutat. Sitkas have also been introduced onto the Kodiak and Afognak Islands, where they thrive.

Like Columbian blacktails, Sitka deer have flourished even after clear-cutting of climax forests and serious forest fires. The deer thrive on the very plants that typically succeed such fires. In fact, California wildlife managers have used controlled fire to improve deer habitat in certain areas where it has become unsuitable. Blacktails actually do their best when about 30 percent of the tall forest in a given area is opened up. Still they need a substantial amount of mature growth for cover and shelter. A sharp population decline always begins where no mature forest at all remains.

My friend Roy Randall, who owns a wilderness lodge for naturalists and fishermen on Afognak, regards Sitkas almost as marine mammals. They are good swimmers and not at all hesitant to hit the cold saltwater along Alaskan coasts to travel to an island or landfall where the forage might be better. Swimming is also an escape mechanism. Mark Mueller, an Alaskan brown bear guide, told me that at dawn one spring morning he saw a pair of Sitka does eating the green sea grass out on the delta of a salmon spawning stream. As he watched the deer from a boat just offshore, he saw a small-to-medium brown bear burst out of shore cover and race toward the pair, whose escape seemed to be cut off.

But as one deer dodged around behind the charging bear, the other plunged into the water where the tide was ebbing. The bruin splashed in right behind her, and at first Mueller thought the bear would catch the blacktail. But gradually the deer pulled away and reached a small islet far offshore. The bear turned back when only halfway to the island.

While salmon fishing with Roy Randall, Peggy and I saw Sitka deer swimming with nothing in pursuit, which Roy did not consider unusual. He said it was not at all unusual either to find bucks traveling over water, particularly during the rut, when the enticing scent of a doe in estrus might waft across a bay or channel on the damp, salty air.

Bears, Blacktails, Pulse Rate

Blacktail deer are by no means dangerous animals, but on a misty morning on Afognak, one buck nearly gave me a heart attack. I had motored by boat from Randall's Lodge to the mouth of a lonely silver stream where coho salmon were beginning to move in great numbers to spawn. These first fresh-run fish are the greatest game species in all Alaska, and I was excited about catching them. Randall had warned me to watch for brown bears, which also were certain to be fishing. I anchored the boat and waded ashore.

Not far from saltwater was a deep, dark pool in which I could see shadows of moving salmon. I cast and at the same time noticed a steaming pile of fresh bear scat almost at my feet. I was looking over my shoulder when a coho struck, catapulted into the air, and spit out my streamer fly all in a split second.

Alaskan coastal brown bears share range with Sitka blacktails but have negligible impact as predators. A bear emerging from hibernation might attempt to take a winter-starved and weakened adult deer, but would be unlikely to catch one otherwise. Occasionally, a bear might happen upon a hiding fawn and take advantage of the fawn's instinct to lie still.

I hooked, played, and released another salmon in that spot. When no more action followed, I began to work my way upstream. The early morning mist thickened into a dense fog, and visibility was poor. Small sounds seemed to be magnified. Some sixth sense caused me to stop again in my tracks, near another pile of fresh bear scat. I listened, and out of sight just upstream I heard a flock of gulls suddenly and noisily flush, a sure sign that a brown bear had chased them. I turned to retreat downstream when suddenly something exploded and came crashing through the alders only a few yards to my right. Another bear, I was sure, only this one had antlers still in the velvet. A blacktail had at first frozen at my approach, but when I came too close, flushed almost in my face. Even after my pulse was finally back to normal, I didn't enjoy fishing that day. Fortunately, I do not have a weak heart.

The Blacktail Ecosystem

A most interesting fact about the various subspecies of mule deer is that somewhere in their range they share territory with at least one of the other horned and antlered animals covered in this book (except for muskox), and usually with more than one. For example, a mule deer living in northwestern Montana might live in the same high-altitude habitat with elk, moose, bighorn sheep, goats, whitetail deer, grizzly and black bears, coyotes, and the occasional wolf drifting down from Canada. There is some conflict and competition among all these animals, but the mule deer somehow manages to survive if not prosper.

It is only with humans that the mule deer and its subspecies encounter serious trouble.

Marijuana and Poaching

Recently blacktail deer have been in the news for an old familiar reason. They have been blamed for devastating valuable crops, only this time the crops are neither grain nor alfalfa. California wildlife officials are suddenly alarmed at the wanton killing of deer by marijuana growers. It is believed that a blacktail can eat eight to ten plants—representing many thousand of dollars in profits—in a single night. In one California area, marijuana growers killed more deer than the legal bag of 2,000 by hunters. One grower killed 25 does and fawns around his plot in one evening. Twelve fresh blacktail hides and heads were later found in his neighbor's barn.

Bob Fox, deputy director of the California Department of Fish and Game, voiced still another worry. In Humboldt County the marijuana growers are also shooting bears, which do not disturb the marijuana crop, on the incorrect theory that strips of bear meat and bear hide hung around a crop patch will repel deer.

The irony is that most of California's estimated 5,000 marijuana growers are producing the high-priced contraband on public lands where it is difficult to apprehend and prosecute them—in conflict with a deer herd not considered all that valuable by most people.

The Blacktail Future

Although not in danger, the continental blacktail population is currently in a slight population decline. I doubt that blacktails will ever enjoy again the same amount of ideal cover available to them in the 1950s and 60s. The large timber companies do not want a lot of deer eating their newly planted seedlings, and they have financed expensive projects to find new and effective blacktail deer repellents. Even worse is the growth of sprawling new communities where deer were once plentiful. These communities often block traditional deer migration routes or over-lap crucial winter range.

One biologist in Oregon pointed out to me that even the four-legged predators are increasing in numbers. Only now they are the dogs of sub-urbanites, which are far more deadly than cougars, coyotes, and wolves ever were. The blacktail is going to need all of its intelligence and coping ability to adapt to the relentless incursions on its habitat.

If this blacktail buck lived another year or two, the third antler tine from the front on each side would have become a "Y" branch, characteristic of all mature mule deer and their blacktail subspecies. (Leonard Lee Rue III photo)

6.

WILD SHEEP

In the lobby of the state office building in Pierre, South Dakota, stands a full mount of an Audubon bighorn sheep. It is a ram, although not a very fine specimen with not quite three-quarter curl to its horns. But it is a priceless display indeed, because no human will ever again see an Audubon alive in the wild or in captivity. The Audubon bighorn has been extinct since about 1924, and that is the only full mount of that bighorn subspecies anywhere in the world.

The Pierre ram was probably killed by an Indian agent, Major James A. McLaughlin, about 1900 near Standing Rock. Herbert Clarkson, a Ludlow rancher who recognized its value and acquired it, donated the sheep to the Department of Game, Fish and Parks. The original mount was in deteriorating condition and had to be completely renovated by taxidermists.

Audubon bighorns were once found in portions of Montana, Wyoming, Nebraska, North and South Dakota, and often in badlands environments. The animals may never have been especially abundant, but there may have been as many as 1,000 still living in 1900. A federal forestry agent reported that 200 Audubons still existed then in the White River badlands of South Dakota, with about 60 or 70 more in the Finger Buttes area of southeastern Montana. In 1958, a rancher near Gates of the Mountains, Montana, showed me the skull and horns of nine rams gathering dust in a decrepit old barn. The rams were killed by his father at Finger Buttes, and they were probably the last Audubon rams in the area.

Unfortunately, the fate of the Audubon sheep may simply be an omen of things to come. There has been an alarming decline in the number of all wild sheep of North America during the past 100 years, despite some great and ingenious efforts to save them. Revered naturalist Ernest Thompson Seton estimated the number of wild sheep between 1½ and 2 million in primitive times, but he believed the United States population was only about 28,000 in 1929. Less than half that many cling to existence south of the Canadian border today.

This Dall ewe and lamb were not alarmed when I approached to photograph them in Denali National Park. Sheep here are fairly accustomed to photographers.

PAGE 126
It is rutting season and this ram, winner of numerous matches against rivals in the herd, is lip-curling in an effort to detect the presence of a ewe in estrus.

Varieties and Characteristics

Dall sheep

Stone sheep

Rocky Mt. bighorn

Desert bighorn

The genus *Ovis* (true sheep) is comprised of several notable forms, including the various mountain sheep of North America; the foreign argali, red sheep, and mouflon; and the many breeds of domestic sheep. Of the wild sheep, both males and females have horns. On females, the horns are modest and goatlike. On males, the horns are massive, typically circling or spiraling, the tips of which may converge or diverge. The curl of the horns may be open or closed. A true sheep's horns may also be large and massive, or thinner and less massive. Horns of American and Siberian sheep are relatively smooth, while those of other Asiatic species have wrinkled surfaces with ripple-like corrugations. The greatest horn development of all the world's sheep is found in the Marco Polo variety of the Pamir Plateau, central Asia, on which the horns frequently exceed 50 inches measured around the curve and where 70-inch curls are not unknown.

Taken as a group, sheep are agile, with a taste for living in lonely, lofty high country. In the Pamirs, their range may extend as high as 20,000 feet, though in Alaska and Canada they live much lower. But sheep are not such rough-country animals as goats, nor are they as good at climbing. Like mule deer, North American bighorn and white sheep are bounding animals rather than climbers, but when threatened they will head directly for the rocks and cliffs. Very steep and rocky places serve as refuges wherever either hunting pressure or predation is heavy.

North American wild sheep are also distinguished from mountain goats by foot and face glands, which goats lack. Male goats possess chin beards and a distinctive odor, while rams are beardless and comparatively odorless.

HABITAT.   With few exceptions, American wild sheep are elusive, located far above the paths beaten by man. Sheep can sometimes be observed at lower elevations and even along busy highways during some seasons, in a few western national parks. But most of the time a hunter, photographer, or sheep watcher will have to make plenty of footprints on ascending trails to find the quarry. The overall impression of an American mountain sheep, especially of a mature male, is of a magnificent and sturdy animal that almost invariably lives on awesome landscapes in undefiled wilderness. No wonder trophy hunters rate the Rocky Mountain bighorn and white Dall with the greatest prizes on earth, and no wonder there is a mystique surrounding the animals. Their scarcity may also have something to do with that.

Early in the 20th century, wildlife scientists used every minor variation in color, size, and general appearance to divide North American wild sheep into as many as 18 different classes. Some of that confusion and disagreement among biologists still exists. For example, the desert sheep (*Ovis canadensis nelsoni*), which we'll cover later in this chapter, is actually a lighter-colored and smaller subspecies of bighorn (*O.c. canadenis*). The Stone sheep (*O. dalli stonei*) is a granite-colored British Columbia subspecies of the Alaska-Yukon white Dall, (*O.d. dalli*), while the so-called Fannin sheep—once recognized as a separate subspecies—is today considered an intergradation between the Dall and the Stone. It lives in a limited area of the Yukon.

To best serve the purposes of this book, I will focus mainly on just two species—the Rocky Mountain bighorn and the Dall.

Early Bighorn Range and Number

Some 19th-century traders, trappers, and early explorers found bighorn sheep to be abundant through the Rocky Mountain chain and westward as far as central Idaho. Captain Bonneville, along with his party and a band of Indians, spent the winter of 1832 along the Salmon River just north of the present site of the town of Salmon. They found plentiful game, including sheep, nearby. According to one account: "Besides numerous gangs of elk, large flocks of the asahta or bighorn, the mountain sheep, were to be seen bounding among the precipices. These simple

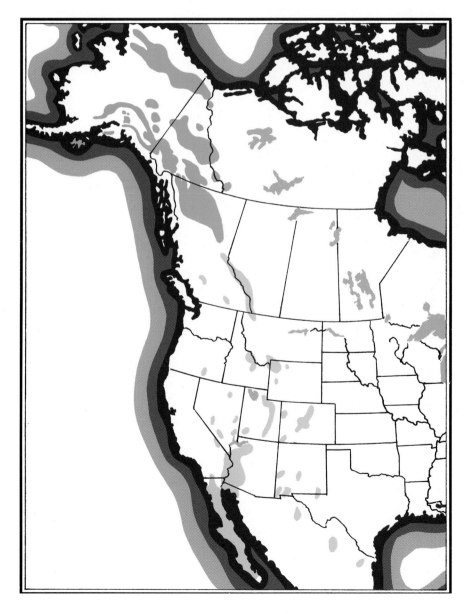

Northernmost is the range of the white Dall sheep; the Stone sheep subspecies of Dall inhabits the southern portion of the Dall range. Bighorns occupy much of the U.S. West and range into Canada, with Rocky Mountain bighorns in the northern portion of the range and desert bighorns occupying the southern portion.

animals were easily circumvented and destroyed. A few hunters may surround a flock and kill as many as they please. Numbers were brought daily into camp, and the flesh of those which were young and fat, was extolled as superior to the finest mutton."

Today, bighorns are only occasionally reported in the same area. Peggy and I have seen scattered bands of sheep while float tripping down the Salmon River. It could be that during the first encounters with white explorers using repeating firearms, bighorns were still relatively unafraid of people. For instance, when the first settlers landed in New England, they were reportedly able to kill the then still tame turkeys, ruffed

Hunting bighorn rams with a camera in the Canadian Rockies has taken us to many wilderness scenes of matchless beauty. No wonder hunting sheep has actually become an addiction.

grouse, and even whitetail deer, standing at very close range. Like the bighorn, all of these today are very wary and wild.

CHANGING HABITAT.   Many other reports of travelers from Colorado westward to Idaho report finding bighorn sheep, even great bands of them numbering into the hundreds in valleys of the Rocky Mountain region. The question is, Were bighorns originally confined to valleys and foothills (as often reported) or did the species also populate the cliffs and steep canyons they inhabit today? We now find much supporting evidence that Rocky Mountain bighorns once roamed far from

the rugged terrain that is now considered their normal habitat. In 1888, for example, Teddy Roosevelt wrote of sheep still being observed on prairies and in valleys many miles from the nearest mountains.

Probably bighorn sheep are simply making their last stand in the mountains, like elk, grizzly bears, and—to a lesser extent—mule deer. Unlike the elk and deer, however, the sheep and grizzlies are not succeeding so well.

Decline of the Bighorn

The general population decline of the Rocky Mountain bighorn began in the 1870s and reached its nadir in the 1920s and 30s. Some herds fared better than others, which were eliminated altogether. Since that low point, sheep populations have known ups and downs but have never really rallied. Most biologists who study sheep are apprehensive, if not fearful, about their future.

Bighorn Physique

Bighorns are much larger animals than they may appear to be when standing at a distance on a mountainside. A large, stocky old ram might

This better-than-average Wyoming bighorn ram was photographed on his winter range. The animal is in prime condition. Unlike horn tips of many bighorn rams, the tips of these horns are not "broomed off." We later saw this ram victorious in several head-to-head duels with other rams of similar size.

stand 40 to 42 inches at the shoulder and weigh 300 pounds. A female might measure 36 inches at the shoulder and average 150 pounds. By comparison, a mature Dall ram will weigh 200 pounds and a comparable ewe about 125.

HORNS. The large and massive horns of the Rocky Mountain bighorn rams continue to grow throughout life. Unlike a deer's antlers, sheep horns are never shed. Rather, like claws, hoofs, and fingernails, they grow from the skin and are composed of a material called keratin, which is entirely different in composition from bony antlers. When plenty of nutritious food is available from the spring until fall, horn growth is steady and comparatively fast. From fall onward until about June, growth is retarded, probably because of changes in body chemistry during the rut as well as the winter shortages in food. The food shortages also cause sheep to use body fat stored during the summer.

The periodic fall-winter slowdown in horn growth results in a pattern of annual growth rings similar to the telltale growth rings of a tree trunk. The pattern appears as slightly deeper constrictions among the corrugations that encircle every sheep horn from the base to the tip. The age of a ram can be determined exactly by the number of constrictions. These are usually so noticeable that a sharp photograph from the side will reveal the age of the subject.

Photographed on Nevada's Desert Game Range in 1972, this desert ram carries horns that are among the largest known for the species.

Behaviors

Peggy and I have spent a lot of time prowling alpine real estate to watch and photograph sheep when we should have been doing something else. But as anyone who tracks sheep soon finds out, it is addictive. You suffer through cold and biting winds, and from aching leg muscles and lungs that almost catch fire, climbing just in the hope that a larger ram is living under the next rimrock, within range of your telephoto lens. If anything is certain, we have learned from our labors that the species is quite unpredictable.

UNPREDICTABLE MOVEMENTS.   Our long field experience with elk and mule deer has shown us that it is possible to anticipate the movements and behavior pattern of those ungulates with some accuracy. But bighorns are consistent in only two habits: taking their middle-of-the-day rest period, and then seeking a familiar or established bedding area each night. Even the noontime rest period is subject to change. On some days a herd of sheep will cease all activities for an hour or two and maybe longer, for as long as they are undisturbed. Most sit motionless, nodding and seeming to doze. But on other days, the noon recess of the same herd is broken when one or more animals stand up, stretch, and feed leisurely for short periods.

MALE AND FEMALE SOCIETIES.   There are really two societies most of the year for bighorn and all North American sheep, and these are sharply divided between males and females. Males band together in smaller groups and live in bachelor associations usually at much higher elevations. Ewes and lambs, in addition to a number of immature males, form their own larger societies, traveling from winter to summer range and back again. When and by what routes they migrate seems mostly determined by weather and by an old dominant female. For instance, the increasing snowfall of autumn, expanding gradually downward and covering food supplies, forces the sheep ahead of it. Males are motivated to move by the oncoming rut as well as by deepening snow. The paths of male and female bands of sheep may cross at any season, but such meetings are not planned and never last very long. The exception to that is during breeding, which takes place in late fall.

Male Bedding Areas

Throughout the summer, rams move between a chain of bedding areas they have established over the years. They rarely graze far away from these areas. Located atop sharp dropoffs and the edges of rimrocks, or just under the crests of ridges, these bedding sites are fairly easy to find, at least into autumn when hunting season opens. Sheep beds are clearly visible shallow depressions, sometimes scraped bare and always surrounded by oval pellet droppings. When hunters suddenly appear in the mountains in September, rams seem to become almost invisible, or at least are always somewhere else. After guiding sheep hunters in areas I thought I knew well, I can guarantee this vanishing act to be true. Rams are most difficult to follow and keep track of in the fall until they suddenly appear in familiar breeding grounds, around November. In Wyoming, the rut takes place while most citizens are sitting down to Thanksgiving dinner. But the rut is worth missing a whole season of banquets to watch.

The Bighorn Rut

Fortunately, there are fairly handy places to observe the bighorn breeding season in Yellowstone and Glacier parks, as well as in all the parks of the Canadian Rocky Mountains. The weather is likely to be cold at best, and the skies slate-colored, but the spectacle of wild sheep in combat is certainly unique and exciting.

Rams may gradually arrive on the rutting site before the females, but not always. One November we watched the ewes for nearly a week, late in November, without a sign of a male. But the next morning we hiked to a meadow overlooking the Yellowstone River, swept bare and hard by incessant and bitter dry winds. There we counted 15 rams of various sizes and ages, from lightweights with only half-curl horns to several old monarchs with heavy, full curls broomed at the tips. We have never seen, before or since, an assemblage of male bighorns to match that one.

That high meadow became an arena for several days of intermittent and sometimes violent jousting. The crash of horns slamming together could be heard half a mile or more away. Yet before the major violence, a lot of body maneuvering, bumping, flank butting, and kicking took place. Then rams of fairly equal size, horn development, and age would square off in their ritualistic duels.

We have been fortunate to watch, although seldom catch on film, the combat at very close range. The rams back apart and, on some imperceptible signal from as far as 35 or 40 feet apart, the two launch into a head-on charge at one another, stiff legs driving as hard as possible. The pile-driver impact may be repeated over and over until one ram is ready to concede, after which he walks or staggers away. From afar the impact has an anvil sound. Close up it is as loud and shattering as the report of a large-bore magnum rifle.

Slow-motion footage has revealed details the eye alone cannot catch. The shock of the horn impact actually ripples through each contestant's body. Dust, saliva, and horn splinters fly. Although blood and gore may drain from noses and eyes, and both rams reel drunkenly from the furious pounding, the duels may continue through dozens of impacts. There are reports of fights lasting for two hours. But most of the bouts we have seen were settled after four or five determined thrusts, and usually fewer.

Years ago in an enclosure of the Desert Game Range in Nevada, we watched a pair of captive desert bighorns in a sharp clash in which one was obviously over-matched. Frustrated, the loser turned and, with its three-quarter-curl horns, easily snapped a ten-inch pine post in two with one lunge.

MATING. Eventually the fighting and bluffing on the breeding grounds sort out the superior rams from the lesser ones, and it has often been written that these winners do all or most of the breeding. Maybe they do. But it is pretty hard to tell for sure in all the confusion. More than one ewe is likely to be in high estrus at any time, and I doubt that the monarch rams can possibly manage all the maneuvering and breeding. (A ewe in estrus is the center of much attention.) Probably the top rams accomplish most of the breeding, but not all.

Early one late November morning, we found a bighorn ewe grazing on the sparse and brittle vegetation of a steep slope above the Yellowstone River, which thundered almost a thousand feet directly below. We stopped for a moment to watch, but mostly to rest tired legs, and that is when four rams appeared from the opposite direction. The leading ram was the largest by far, with horns heavy enough to excite any trophy hunter. The other rams followed in order of descending horn size, the smallest being last. All the rams knew immediately that they had found a female coming into estrus. The lead ram approached with head and neck lowered from the shoulders, snout raised slightly and head turned to one side to best show off the full, thick curls of his horns. The ram's upper lip was curled back in the manner of other big-game animals when the scent of estrus is in the air.

Without even looking up from grazing, the ewe suddenly started uphill on a dead run with the rams right behind. The animals moved so fast that we could not follow them easily, but in a few seconds we heard the metallic crash of horns. The pursuing ram had paused just long

Having won the ritual battles for top rank among males, this Wyoming ram follows a ewe to a secluded steep mountainside where breeding will take place. The ram will then try to find other receptive females.

PAGE 136
During November in Wyoming, bighorn rams join females already on traditional mating grounds, and the ritual that will determine rank and stud status begins. For days, maybe weeks, rams strut and posture, challenge and parry, all culminating in the sharp clashes that characterize the battles between all wild sheep rivals. Ordinarily rams fight "by the rules," butting head-to-head. But occasionally a maverick will slam a rival in the flanks or rear. It is a rough show, exciting to watch.

enough to keep his nearest rival at a distance. We've seen this happen before, and almost always these running clashes are of short duration. Occasionally the ewe stops to graze. Most often she is driven to keep moving.

One such fight we witnessed was nothing like the classic head-to-head duels of a week earlier. During this peak of sexual excitement, one ram might strike another without warning in the flank or rump or any other exposed place. We saw one ram smash into the body of another that had lost his footing, sending him rolling over a ledge and down a rocky slide. The blow and fall would have killed any person, but in a few minutes the animal was back in the chase and apparently none the worse for his experience. Sometimes when two rams pause in a chase even briefly to fight, another ram is able to dart in and mount the ewe. Or, a ewe might be covered by two or more rams in the space of a few minutes. As I said, it is all very confusing.

Birth and Development

Somehow, all of the ewes are bred by the end of January, and by mid-June in the northern Rockies—about 180 days after the rut—the lambing period begins. The same lambing areas are used year after year, and they are usually in the most precipitous and inaccessible parts of the herd's range. With only rare exceptions, ewes have single lambs that stay with them in rocky, remote places for two weeks or so. After that, females with lambs reassemble into small bands.

A bighorn (or a Dall) lamb is precocial and grows rapidly during its first summer. By the next rutting season in late fall, it will be half the size of its mother. Peggy and I have seen lambs only a few weeks old already eating a variety of fresh green vegetation. They soon are very agile and surefooted and gambol all summer long with one another. Lambs especially relish playing and running on the lingering snowbanks in their summer range.

BELOW LEFT
A very young Dall ram, just over a year old, approaches for a better look into the business end of a telephoto lens in Alaska.

BELOW RIGHT
Although extremely shy beyond park boundaries, this bighorn ewe and lamb pose beside a Yellowstone Park highway over Dunraven Pass, a bonanza for park visitors with cameras.

As summer wanes, ewes seem to grow less and less interested in their lambs—and less protective and tolerant of them. By this time the lambs have a long-haired, raffish appearance. By September we have seen lambs trying to suckle that were emphatically turned away. Once or twice a lamb that was too persistent was butted and sent rolling off its feet.

Life Expectancy

Biologists who have carefully studied sheep report that lambs can survive most winters without their mothers if they are not separated from their own bands. In fact, surviving its first winter will only be a lamb's first and not necessarily greatest obstacle to living out a normal life. In areas where there is no hunting at all, or where only large rams are hunted, statistics from several states and Canadian provinces suggest that a bighorn has a life expectancy of about four years at birth. Keep in mind that this figure also includes lamb mortality, which can be high.

Predators

From the time a sheep is born deep in the mountains, it has many enemies, animal and viral, as well as human. The animal predators are likely to seem the most dramatic and serious, but they are not. For example, coyotes may be numerous over much sheep range, but big-game biologists agree they are a minor contributor to herd drain.

Considering its size, strength, and habit of hunting in rough country, the mountain lion is potentially the most serious bighorn predator. In certain areas lions do take a toll, which has been especially severe on relocated bighorns—animals newly introduced to repopulate an area where sheep were wiped out. Established sheep, as well as other big-game species, have a definite advantage over recent transplants against predation because they know the terrain better.

In Alaska and Canada, wolves periodically threaten Dall and Stone sheep herds, especially when the deep snows of late winter and early spring give these predators a hunting edge. Golden eagles are often cited as sheep nemeses, and so are carnivores from grizzlies to wolverines and even lynx. But all of those lumped together are insignificant compared to man.

Poaching

Poaching for trophy rams has become almost a way of life wherever there are rams with bragging-size horns. A set of record-book horns is valued so high that the poaching season never really closes. In many sections of the Rockies, only a fraction of the rams die of old age.

In March 1983, an eastern big-game hunter arrived at the Missoula, Montana, airport in a snowstorm. He was met by two men, one a Denver police detective, who would guide him on a hunt for bighorn sheep in the Glacier National Park area. The hunter had no Montana hunting license and the hunt was strictly illegal. The two waiting men, detective Dale Leonard and his partner, had guaranteed their "client" a record ram for $7,500. The fee was fairly low considering the serious risk and high value of such a rare trophy.

What the illicit guides did not know was that their client hunter was a federal undercover agent. Evidence from this "guaranteed hunt" and from other similar busts helped crack a two-state commercial hunting ring that dealt in trophy deer, elk, and bears as well as bighorn sheep. Leonard pleaded guilty to two felony counts. Six other members of the ring were also found guilty and fined. Some were given a jail sentence. But that case may only have made other poachers more cautious.

And no wonder. According to U.S. Fish and Wildlife Service sources, an outstanding bighorn sheep trophy may bring as much as $20,000 in the United States, Europe, or Japan. More and more illegal trophies are turning up in collections overseas. The coveted Grand Slam —which consists of the horns and cape from a bighorn, a desert big-horn, a Dall, and a Stone ram—can sell for as much as $50,000.

Here's how a typical sheep-poaching operation works. The poachers scout areas where trophy rams live, especially wintering areas, which are more accessible and often well known. They may even photograph the animals and send pictures to interested parties, specifying a price for the head. The clients can then visit the Rockies and shoot the animal for themselves or simply have the poachers ship it to them. One anonymous person called and even contacted me a second time to ask where I had taken the photo of a certain bighorn sheep published in *Outdoor Life* magazine. Probably he was a member of a poaching group looking for a new source of sheep horns. He didn't get the location from me.

Clever poachers benefit from the fact that in most western states it is legal to sell horns and antlers legally taken by licensed hunters. None-theless, a three-year sting operation conducted by the U.S. Fish and Wildlife Service to apprehend a crime ring of hunters, outfitters, and taxidermists resulted in March 1985 in what was the stiffest sentence ever handed down in the country for a game violation. A U.S. District Judge sentenced a Montana man to fifteen years in prison for trading in illegal trophies. Three other defendants in the same ring received heavy sentences for conducting an illegal hunt and selling illegal bighorn heads.

The sting cost over $200,000 and was worth it. Yet it is sad that this money could not have been spent on acquiring critical winter range for sheep, or for an important scientific study.

Diseases

Still, like natural predation, poaching is only a dramatic and easy expla-nation for sheep decline, one that makes headlines. Lungworms, which parasitize all species of North American sheep, are far more deadly. So are a host of other maladies such as bluetongue and pinkeye, soremouth and pneumonia, mange (spread by a mite), and sinusitis, which begins with a nasal botfly. Pinkeye has come close to decimating a bighorn

It is springtime and this Montana bighorn ram has survived another bitter winter, though just barely. Emaciated and weak, he will recover some vitality and grow fat during the coming summer, but the next winter may well be his last.

herd in northern Yellowstone Park since 1982. During 1983, in Waterton National Park, Alberta, pinkeye reduced the population of sheep to a third of its former size. The bighorn herd in Hells Canyon, Idaho, is not healthy today, either.

Why are our wild sheep so susceptible to those diseases, which they apparently did not have in primitive times? Most of the answer lies with domestic livestock, particularly domestic sheep. These sheep not only transmit diseases to which they are comparatively immune to wild sheep, but they also annually overgraze an incalculable amount of wild-life habitat in America. Government agencies, which permit this, are thus sacrificing precious national wildlife for a dying industry—sheep ranching.

In 1971, the California Fish and Game Department released 10 healthy bighorns from British Columbia into an 1,100-acre fenced-in area of prime habitat in the Lava Beds National Monument. The area was adjacent to Modoc National Forest. The new herd was thriving and grew to 42 until the Forest Service permitted domestic sheep to graze in the Modoc right next to the enclosure. By 1975, the wild sheep began to die of soremouth. A veterinarian, Ted Kistner of Oregon State University, warned that the bighorn program was threatened by the domestics.

In mid-1980, after the Forest Service agreed to a half-mile buffer zone along the fence, a sheep rancher moved his flock into that zone anyway, and Forest Service officials refused to evict his herd. One morning Lava Beds rangers witnessed nose-to-nose contact between domestic and wild sheep right along the boundary fence. The next bighorn died a week or so later, and by August 1980 all had succumbed to pneumonia.

Protectors of the Bighorn

Some dedicated people, both laymen and professionals, go to great lengths to conserve bighorn sheep. Safari Club International, a sportsmen's group based in Tucson, Arizona, regularly donates money to survival measures and study projects for various sheep and other big game. A new Society for the Conservation of Bighorn Sheep has been formed,

and is based in Cody, Wyoming. The Sierra Club and the Wilderness Society have long been engaged in saving from development many wild areas where sheep might continue to live in North America. The National Wildlife Federation, the National Audubon Society, and the National Parks and Conservation Association have also long been involved in both programs and a general philosophy to benefit sheep. The Nature Conservancy has even acquired scattered lands where sheep and other wildlife can live free of domestic creatures.

PAGE 142
With astonishing capacity to endure heat and dehydration, the desert bighorn is the "camel" of North America. Rams have been known to fight in 113°F heat for two hours and suffer no apparent ill effects.

The Desert Bighorn

The desert subspecies of the Rocky Mountain bighorn is truly a remarkable creature in its own right. Desert bighorns—*Ovis canadensis nelsoni* of California and Arizona, and *Ovis canadensis mexicana* of Mexico—have a tremendous tolerance to extreme heat and can go for extended periods without drinking at all. Researchers in Death Valley National Monument, California, have observed a ram chasing a ewe at full speed during a day when the official temperature was 122° F. On another occasion two large rams fought near, and perhaps for possession of, a waterhole for two hours nonstop in 113° heat, apparently with no ill effect.

Among large mammals, only dromedary camels of the Sahara Desert can equal and probably surpass the desert bighorn in its ability to withstand serious dehydration. These sheep may be able to lose almost one-quarter of their body weight without being critically weakened. Biologists have often described how desert bighorns arrive at waterholes emaciated and seemingly near death, with eyes dull and ribs and hip bones protruding. But then their recovery is almost miraculous. After drinking until their shrunken stomachs bulge, and after a brief rest in the shade, the same sheep would walk nimbly away through the heat waves of the most desolate country on the continent.

Yet for all of its tenacity, the desert bighorn is also the most vulnerable of American wild sheep. Because record-class horns from this animal are the rarest of those within a Grand Slam, or Big Four collection, authorities in California have had to maintain an almost constant vigil over scattered desert sheep populations. California biologists have discovered also that many of these sheep do indeed die of thirst in the extremely dry, harsh desert environment.

In 1969, California wildlife personnel found a sheep death trap that had claimed at least 34 animals in the Chocolate Mountains. The trap was actually a natural pit 10 feet deep and about 12 feet in diameter etched out of solid rock. Nature's bait for the trap was runoff water that filled the pit after every desert cloudburst, and then stayed there while other water sources dried up. Sheep could drink easily from the pit when it was full. But when the water level dropped, sheep fell, or possibly jumped, into the trap for a drink and could not get out.

Fortunately, the U.S. Marine Corps lent a helicopter to carry rock drilling and blasting equipment to the site, which showed no evidence of any previous human intrusion. California wildlife personnel then blasted an escape route out of the pit to provide easy escape for many creatures, no matter how low the water level. While they worked, the temperature rose well above 110° F, as it often does in this furnace environment.

Because the difference between survival and death comes down to a year-round supply of water for many sheep in Mexico and southwestern United States, some remote populations in widely scattered mountain ranges have been given a lease on life by the building of artificial guzzlers and enlarging or improvements of existing seeps.

The Dall

It is interesting to note that the northernmost of North America's wild sheep, the Dall (*Ovis dalli dalli*), is also a product of relatively dry country. That may be hard to believe for a visitor to Denali National Park, Alaska, who hopes to see sheep but rarely does because of low-lying mist and a sometimes steady summertime drizzle, which is so common in that part of the Alaska Range.

HABITAT AND FEEDING HABITS.   The Dall thrives best in a combination of open alpine ridges and meadows with steep slopes, cliffs, or other rugged escape terrain in the immediate vicinity. Their feeding and loafing areas must always be within reach of a safe sanctuary from predators. Unlike Rocky Mountain bighorns, which may occasionally venture down into evergreen timber to feed or to visit mineral licks, Dalls rarely venture below timberline, except when a band moves very

We found this fine Dall ram in Kluane National Park, Yukon Territory. He has very good, heavy horns for the species, and was roaming not far from the Alaska Highway.

quickly and directly from one range of mountains to another. The rams live in bachelor groups, separate from the rest except during the annual rut.

These white sheep feed primarily on grasses, leafy ground plants, mosses, and lichens found on open mountainsides and on ridges. In spring the herds generally follow snowlines upward, taking the reverse path in autumn. Time and again we have seen sleek and healthy sheep living on hillsides with very little vegetation and none of it growing more than a few inches high.

Dall sheep are able to dig down through a foot or so of loose snow to reach food. But deeper packed snow or icing conditions can cause Dalls to move to less secure range where they are more susceptible to predators or starvation. But according to naturalist Adolph Murie— brother of Olaus Murie, mentioned earlier—who spent many years observing both Dall sheep and wolves in what is now Denali National Park, sheep in good health are a match for wolves or any other predators.

In early summer, Dall sheep are scruffy in appearance as they shed the previous year's white coats. The old coats are quickly replaced with sleek new, thick growth that allows the species to withstand long subzero winters.

HORNS.   Dall sheep inhabit high ranges of Alaska, Yukon, northern British Columbia, and the western edge of Mackenzie District, Northwest Territories. This species grows thinner horns than the bighorns to the south. Older Dall rams have long, widely curving horns while ewes and young rams have short, slender, slightly curled horns. The horns of a ram of Alaska's Wrangell Mountains will not reach a full curl until the animal's eighth or ninth year.

AS CAMERA SUBJECTS.   One recent September, Peggy and I were traveling southward on the Alaskan Highway after spending a golden summer in the Great Land. Three weeks of that time had been spent in Denali, which is among the best, most accessible places to see the monarchs of the Alaskan wilderness. We had focused on everything from wolves and grizzlies to barren-ground caribou and moose. But despite our many days of fruitless climbing to find them, somehow the white sheep had eluded us. Now we were driving home through Kluane National Park, revelling in the exquisite, tawny Yukon scenery and fall weather, and not thinking about sheep at all. Suddenly we saw two all-white sheep standing just above the gravel washboard roadway, and I braked to a halt.

Less than 100 feet away a pair of Dall rams were standing and watching us. On one of these, a horn was broken off. But the other horn was a splendid full-curl trophy.

With unsteady hands and heart pounding, I grabbed a camera, affixed a telephoto lens, and extended the legs of a tripod in a desperate race to shoot the animals before they charged up the steep mountainside just behind them. But I need not have tested my blood pressure. The rams simply stood and grazed on coarse grass, pausing occasionally to look at us. I've never found any wild sheep easier to photograph than this duo in Kluane.

That leads to the greatest mystery, at least to me, about wild sheep. Most hunters agree that they are among the wariest, most difficult of all large wild animals to approach. And I agree. But whenever they are not hunted for a certain period of time, as in Kluane, they soon become far more tolerant of people than are deer or elk. I don't know why this is so.

7.
MOUNTAIN GOAT

A little over two centuries ago, as he was exploring northward along the fiord coast of British Columbia on a voyage around the world, Captain James Cook dropped anchor near a forest of totem poles on the beach, indicating an Indian village. Once ashore, Cook noticed the soft, white, spun-wool garments worn by some of the inhabitants. He asked about the source of the wool, and the Indians pointed to all-white animals high on steep cliffs looming behind the village.

We know now that Cook could never have seen the animals close up, because he later wrote of them as "polar bears." In reality they were mountain goats. Still, people have made similar mistakes since then. Most often the white goats are called sheep, or vice versa. Goats are nonetheless among the most interesting horned animals.

For the record, the mountain or white goat, *Oreamnos americanus*, is not a true goat but, rather, is related to the chamois of Europe but not to any native big-game species. That Latin name, incidentally, comes for the word *Oread*—or mountain nymph in Greek mythology—which aptly describes the animal's unique adaptation to mountain ranges of northwestern North America. The mountain goat is also similar to the tahr and serow of the Himalayas. I have seen the tahr and, except for its heavier horns, that animal at first appeared to be a darker version of the American goat. Henry Alexander, a now-obscure naturalist, was the first to accurately describe and name the mountain goat.

Range and Habitat

The original range of the white goat was fairly well defined, including most ranges of the northern Pacific coast from Washington to southern

From a very early age, kids are able to follow their mothers wherever they travel, over any kind of terrain. Life itself depends on never making a misstep, no matter how precarious the footing becomes.

PAGE 146
Peggy and I were resting one day after a long climb when we were suddenly no longer alone. This billy appeared, studied us, then bedded down not far away, keeping an eye on us.

Alaska, and inland to the British Columbia Cassiars. Also included were the northern Rockies of Idaho, Montana, and Alberta. White goats have been stocked as far east as Colorado and the South Dakota Black Hills, where the small herd seems as unnatural on the landscape as Mt. Rushmore. Another herd lives in the Beartooth Mountains northeast of Yellowstone Park. But no matter what the location, original range or not, the animals live their lives largely in the sheerest, least accessible places they can find, often above the clouds at altitudes up to 10,000 feet.

Our own frequent experiences with goats have ranged from extremely pleasant to hair-raising. The latter happened on one of those days when I eagerly climbed higher and higher into goat country to take pictures, only to later sit weak and nearly frozen, wondering how I would get back down. In that respect, goats are dangerous. They can lure you into predicaments where even technical mountain climbers should proceed cautiously.

Yet it is that same edge-of-the-world, edge-of-life habitat that has permitted mountain goats to escape predation from wolves, grizzlies, cougars, and, to some extent, man. It has also enabled goats to avoid the investigations of scientists. Only the relatively recent combination of helicopters, ultralight climbing equipment, and biotelemetry has permitted biologists to probe into the lives and travels of mountain goats. Suddenly the mountain goat has emerged from legend into a more familiar symbol of the wilderness.

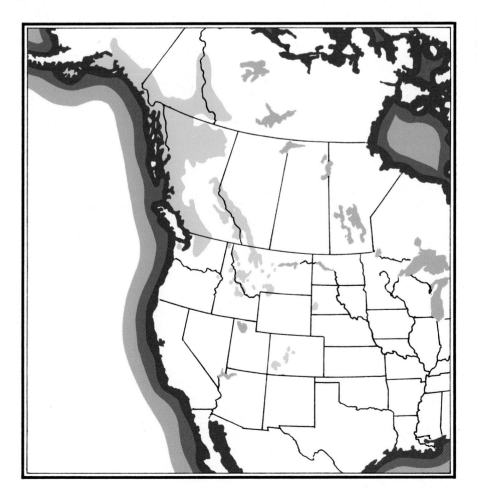

Mountain goats survive in scattered herds in high terrain throughout the West.

This shedding goat lives in the Beartooth Mountains, a vast wilderness northeast of Yellowstone Park. At times the goats can be seen from the Beartooth Highway, which is one of the highest paved roads on earth.

The Physical Specimen

The mountain goat is an impressive and powerful animal, usually much larger at close range than it appears on a distant mountainside. Look at one dead-on, and the impression may be of a goateed old man of the mountains. But there is much more than meets the eye.

WEIGHT AND SIZE. Among the earliest serious investigations of white goats were those by Stewart Brandborg in Idaho. He was able to weigh a number of animals, and he found that yearlings averaged about 42 pounds, with males and females running about the same. The animals gained 20 to 25 pounds by the end of their second year and weighed well over 100 pounds when three years old. The heaviest animal weighed by Brandborg was a billy of over four years and 194 pounds, which measured 5½ feet long and 3½ feet at the shoulder. But 200-pound males are not uncommon, and a few probably exceed a live weight of 255 pounds. Goats of the same age vary greatly in size, depending at least to some extent on the availability of high-quality foods on range that is not overpopulated by the herd.

PROTECTIVE COAT. The coarse guard hairs that grow as long as 8 inches along the back and down the legs give the animal its characteristic forward-leaning profile. The long chin whiskers are responsible for the name goat, which it superficially resembles. Unlike domestic goats and some wild ones, female mountain goats exude almost no odor what-

ever, although billies do have a characteristic smell on damp days. The dense underfur is very fine and of a quality that compares to fine cashmere. With the help of their pelage, goats can withstand both bitterly cold and very damp winters if they begin the season in relatively good condition.

HOOFS. The hoofs as much as the coats make life possible in typical goat country. Not long after birth, a kid displays as great dexterity on rough terrain as any full-grown American horned or antlered animal, including the wild sheep. Instead of the concave surface within the horny shell of the hoof, as in deer, a white goat has pliable pads that are slightly convex so as to protrude beyond an outer hard covering. These provide the reliable grip a goat needs to either gambol, as young ones do, or plod deliberately and confidently as does a large billy. On soft ground the squarer shape of a goat print permits a tracker to distinguish it from that of a bighorn sheep or a mule deer, the only two ungulates likely to be found in the same range.

DELIBERATE REACTIONS. I have heard many mountain people in goat country (including some with long experience at goat watching) say that the senses of the animal are not very keen. That misconception probably results from the animal's deliberateness, its attitude, and from the fact that it does not react or recoil instantly at the approach to danger as does a whitetail. Only rarely does a goat *have* to react immediately. Safe refuge may be only a thin ledge a hundred feet

Photographed in early summer, members of the same Beartooth herd as the goat on the previous page are shedding their old coats.

away, where any predator that follows will have to face a pair of stiletto horns on the edge of space, with no room to retreat.

EYESIGHT AND HEARING.   Studies have revealed that a bedded goat is aware of anything moving as far as a mile away. A goat is quick to detect movement, but may be puzzled by a stationary object. The species also has very good hearing. Mountain goats can instantly distinguish natural sounds, such as those of rocks dislodged by freezing and thawing, from those made by something approaching unseen. Without seeing the approaching animal, a goat also can tell another goat (and maybe even *which* goat in its herd) from, say, a mule deer or sheep. The goats in the accompanying photos did not climb away when we approached to photograph them because they knew very well there was nothing to fear. In many areas where they have not been hunted over a period of time, goats (like bighorn sheep) become quite tolerant of photographers, which is another reason I like them.

Horns—Male and Female

Both male and female mountain goats have the same kind of short black horns, the male's averaging larger and longer. Although small compared to those of other large mammals, mountain goat horns are important for defense and for maintaining dominance or rank throughout their lives. These are true horns, never shed, and the growth rings provide a permanent record of each animal's living conditions. There are two parts to the horn. First is the core, an outgrowth of the frontal bone of the skull. What we see is the core's hard, black, sheathlike covering. Epidermal in origin, the sheath grows out from its base. After a goat dies, the sheath separates easily from the core.

ESTIMATING AGE.   Most of a white goat's horn growth is achieved by the time it is four years old. Beyond that the length increases very slowly. Age of the animals can be determined by counting the annual growth rings, though younger animals are also aged by dentition, as are deer. Yet is very difficult to determine the age or even the sex of a mountain goat without stalking very close to it.

Survival of the Fittest

Many large mammals can recover from heavy hunting pressure and predation by producing more young or with a higher survival rate among the young born every spring. But studies have shown this is not true with mountain goats. Many goat herds will maintain their own balanced populations as weaker animals starve or are butted over the edges of cliffs during daily skirmishing. More than once, Peggy and I have seen larger goats try to force smaller ones out onto thin ledges from which there was no return. Orphaned kids—or malnourished ones—are especially likely to be pushed over a precipice if they happen to be in the way of a larger, stronger animal seeking food or even a favored bedding

site. More often an abandoned or weakened kid is simply driven toward colder, more barren places where little food exists.

A pair of billies stroll along an Olympic National Park ridge that is swept by ocean winds and often shrouded in mist. A long, lung-busting climb is required before you can photograph them.

Feeding Habits

When introduced into mountain ranges where they did not exist originally, goats can become as destructive to fragile alpine vegetation as they are to one another. In 1920, about 15 goats were released from the Cascades into Washington's Olympic National Park. In 1982, the population was estimated at about 500, far more than the high country vegetation there could stand. The animals had caused such devastation to plants that a program to live-trap and remove most of them was begun in 1984.

Winter Range

We do not usually regard goats as migratory creatures, though they do travel regularly with the seasons. Their travels tend to be down from summer range and up again in winter, rather than laterally for great distances like those of caribou, elk, or bison. Some herds may have to make a fairly long trip, however, to find the right combination of steep broken terrain and available food to survive the winter. Most wintering areas are on south-facing slopes because the snow will not be as deep there. Places blown free of snow by wind are also favored.

One goat or even a herd of them might manage to find enough sustenance on windswept, overgrazed slopes during summer; but food gets mighty scarce when the first snowfalls of autumn reclaim the ridges, and forage is hidden beneath deep drifts. The same ledges where footing was firm (at least for a goat) are all at once crusted with snow or, worse, glazed with ice.

Winter's Toll

I've often read how golden eagles, for example, take a toll of white kids by diving and forcing them to plunge off of cliffs to death below. That may even happen on some rare occasion. But despite all the climbing skills acquired during a summer of serious play and practice, winter is a more dire threat than predation for any young goat—or adult, for that matter. Much of American goat country, especially the coastal areas, is also a high precipitation belt of the Northwest, where a single storm front can drop 4 feet of wet snow overnight. Without the fat reserves and larger heat-retaining bodies of adult goats, kids must use more energy just to keep warm. If it is to be among the 50 percent or less that survive until spring, a young goat needs a dominant-female mother to break trail, find sufficient food under the snow, and then protect the food from others in the herd.

The danger of avalanches is ever present. Avalanches can occur at any time of the year, and range from minor snowslides to devastating cataclysms in which whole mountainsides fall away. But the worst of them take place mostly from midwinter, when layers of snow begin to pile up, crust, and shift, until early spring, when there is much alternate freezing and thawing in the high country.

An avalanche can originate as nothing more than a few rocks being dislodged by ice thawing. But it can gain both material and momentum with astonishing speed, and from firsthand, close-range experience I know it can be a terrifying spectacle to watch. I shudder to think what it would be like to be caught in the path of an avalanche or be carried inexorably downhill in the deluge. Every winter a substantial number of large mammals become avalanche casualties, but goat losses may be the greatest.

After six years of studying goat carcasses found in early summer at the bottoms of avalanches, Glacier National Park biologist Douglas Chadwick watched a nanny swept downward by an avalanche that the animal may have herself triggered. He was lucky not to be caught in it himself.

Chadwick was cross-country skiing through the park on a glittering bright April day when he spotted the goat, which was familiar by the unusual shape of her horn. The biologist wanted to see if the goat was pregnant, but wet snow was balling up on the underside of his skis. He paused to change the wax on the skis, and that may have saved his life. Just ahead on the trail, the mountainside gave way and became a deafening thunder of rocks and ice. For a time the nanny managed to stay atop the avalanche, as if riding it out to the bottom. Then the stunned Chadwick saw her vanish.

Thad Mann, a British Columbia fishing guide and naturalist described to me a similar experience. He was hiking on a spring morning following a clear, cold, salmon-spawning stream. As he recalled, the white animals were fairly well scattered over the cliff when the center section began to fall away, at first as if in slow motion, then gaining speed. Five goats standing in the center of the herd simply disappeared. Much later Mann investigated the small mountain of debris where the avalanche came to rest, hoping for some sign of the goats, but there was no trace of them.

The end of winter often finds goats at their lowest elevations of the year. The early-bird river rafters on Idaho's Salmon River, the River of No Return, sometimes see them on the canyon walls just above the water. In springtime I have also seen goats from a canoe along the Missouri River at Gates of the Mountains, Montana. But all across their range today, the white goats are able to cope with winter at higher eleva-

Mountain goats seem programmed to butt each other year-round, whether in dispute over a feeding spot, over rights of passage, during the rut, or—especially among kids—for sport.

tions and in harsher conditions than mule deer, elk, and bighorn sheep, which share their habitat during better times.

Butting

During the fall and early winter of 1985, Peggy and I spent more time in goat country than ever before. We photographed the animals in a variety of situations and places, from Montana to the Yukon Territory at the northern limit of their range. Perhaps our clearest impression, next to their choice of living in a perilous environment, is that mountain goats are programmed by Nature to butt one another at every opportunity. That is equally true when a nanny meets another nanny in her path or a kid not her own, or when a male meets another male or a female, any time at all. It occurred to me that butting may even be a means of recognition or keeping in touch. But butting takes on a new vigor and significance during the rut, which may occur from mid-October to December, depending on the location. Breeding usually begins earlier in coastal Alaska and British Columbia than inland in the Idaho and Montana Rockies.

Fatal Clashes

As with other large mammals, most clashes of rutting males are settled by horn threats, posturing, humps raised, or aggressive rushes, from which one goat retreats. But in 1927 Ernest Thompson Seton, among other later observers, reported fights that were fatal. On one occasion after a sharp headlong attack toward the body, I saw a male butted from a deadend ledge. He fell about 12 feet into a jumble of jagged rock where he should have been seriously injured. But the animal immediately climbed back toward the scene of all the rutting action, apparently unhurt. He wisely avoided the billy that had pushed him over the ledge.

Mating

One of the better-known legends concerns the sex life of goats. After a summer hunt in Montana in 1912, the old hunter and conservationist Teddy Roosevelt shot a billy, nanny, and kid that happened to be together at the time. From that experience he concluded that the species was monogamous, a belief I heard repeated by a Yukon Indian seventy years later.

The truth is that goats are highly polygamous, which is why billies of all ages spend the rut in constant search for estrus females. Once a female is found, a large male follows her anywhere her trail or aura goes. By butting or threatening to butt, he drives lesser males away and in time may himself be driven away by a stronger billy. We have watched billies force nannies from their beds and begin to butt and bump them with horns and body, only to turn, scale a cliff, and repeat the process on another female bedded down 200 yards directly above.

One day we watched one male in a positive frenzy, commuting back and forth until nightfall between what seemed to be two equally attractive females on separate rock ledges. En route in either direction, the billy had to threaten another billy that had similar interests. We have often wondered how the strange tryst ended, because by next morning the herd had moved just far enough to be out of good viewing range. The species, incidentally, does not tend to gather into larger herds during the rut.

Birth and Development

Toward the end of May, pregnant females withdraw from their small herds and climb to secluded ridges and crevices, sometimes on the faces of cliffs. In such dangerous retreats, single pure-white kids weighing

from six to seven pounds are dropped into the cold, thin atmosphere. For an hour or so the kid may stand somewhat unsteadily and try to nurse. But no native mammal babies are more precocious than these. According to biologists who have witnessed the birth of a baby mountain goat, before one day old they are able to climb with apparent confidence where no sensible human would try to follow.

Kids are born with innate climbing skills, if not a total lack of fear. Their flexible hoof pads allow excellent traction. In Glacier and Olympic parks, we have watched very young ones scramble along the edge of eternity, jumping from loose rock to ledge and back again and leaping over and onto their patient mothers, in an apparently reckless way. Still, the survival of kids matches or surpasses that of such other big-game species as mule and whitetail deer, elk, and pronghorn.

Babyhood is very short in many ways. I do not know how long kids continue to nurse, but we have seen week-old young nipping at the same plants on which the mothers fed right next to them. Little ones, especially males, begin to practice sexual and fighting postures at an early age, running and then suddenly whirling about to face an imaginary rival. When a nanny and new kid rejoin their herd within a week or two, play and play fighting with other new arrivals can become very spirited and very rough. We have often seen kids deliberately try to push "playmates" off of such steep cliffs that a fall would have resulted in certain death. The same mothers that jealously guard themselves and their kids from an invasion of personal space or feeding area seem

Come winter, this kid will depend heavily on its mother to break trail in the snow and scrape down to food under the snow. Often, 50 percent of the goat kids fail to survive the winter.

Goats do not share their mountain habitat with as many different creatures as, say, a whitetail deer. But one mammal a goat photographer is likely to encounter, or at least hear whistling in alarm, is the hoary marmot.

unconcerned that their kids spend hours testing one another on the brink of nowhere. Occasionally one of the less strong or surefooted is pushed over. Once more, it is a case of survival of the fittest.

Although females do not normally reach sexual maturity until 2½ years old, we have seen males following smaller females that appeared to be only 1½ years old. We have also seen very young males following much larger females, perhaps seriously or maybe just for the practice. One of these precocious males was suddenly given a sudden hard butt for his attentions, but he wasn't really discouraged.

Capricious Behavior

I live 48 hours every summer day that I am in goat country. For one thing, I've learned that the same animals that may at times seem lethargic can also be very lively. Sit down in a comfortable, convenient place to watch a herd of nannies with half-grown kids and pretty soon fights are breaking out all over. The squabbles are for dominance, space, and choice vegetation. The whole mountain may seem to be in turmoil when suddenly, as if on an unseen signal, all bed down and everything is peaceful for a period, usually at mid-day. Only a few kids may keep charging about in youthful high spirits.

Before daybreak one morning, Peggy and I began climbing along a popular hiking trail in Glacier Park. This is a good trail for hikers interested in goats if they have stamina enough to hike toward Gunsight Pass. But after we'd tramped steadily for three hours until well past sunup, the only creatures we had seen were a circling golden eagle and a hoary marmot, which whistled at us as we passed. We sat down on the top of a sharp drop-off and I began to melt snow for coffee water on a small backpack stove. That's when we heard loose rocks rolling out of sight in a mini-avalanche directly beneath us.

A moment later the unseen rock-roller, a young billy, climbed up from just below and bedded down about 20 feet away. Although I'm sure he noticed us, the animal gave little indication of it. The billy was still there when we had drained the coffee, shot several exposures of our new friend, and continued our hike. We concluded that the goats we meet along the trails are what make Glacier an especially stupendous national park.

Fatal Curiosity?

It is often written that goats require totally undisturbed wilderness to survive, but that simply is not true. I personally do not want to see another square inch of the American wilderness lost for any reason whatsoever, and in fact believe that far more land should be set aside as wilderness. Yet I know of too many places where goats do not bother to avoid people, as well as their cars and dwellings, even when it would be easy to do so. I suppose if they did, though, there would be far fewer goat photographs in photographers' files.

Neither roads not the presence of people drive goats from their selected habitats. But poaching—often made easier by the building of wilderness roads—can be damaging to a herd. The problem begins when these naturally curious animals pause too long to watch a person moving toward them. It is then that they are an easy mark for flat-trajectory, high-powered rifles in unscrupulous hands.

Stalked by Wolves

In the early 1950s, my friend Frank Sayers and I had an unusual opportunity to watch a goat's calm reaction to danger. We were bivouacked beside Tuaton Lake in the British Columbia Cassiars and had just returned to camp with enough rainbow trout for dinner when somebody noticed several white objects on the mountain behind us. Through a 32× spotting scope the white spots became four nannies, each with a single kid, bedded down on what seemed to be narrow ledges. After studying them for a few minutes, Frank suddenly whistled softly.

"Look just underneath the lowest pair," he said.

I focused and immediately saw two wolves stalking upward from directly below the goats, and already not very far away from them. From our long range and perspective at least, it seemed that the goats were unaware that any danger existed any nearer than us fishermen camped on the lakeshore. I was certain we would soon see the wolves capture the kids. Suspense began to build as we watched the wolves creep closer and closer.

But the stalk turned out to be an exercise in futility. When the wolves were only a hundred feet or less away, the nanny stood up slowly, stretched, and began climbing directly upward with no sign of alarm. The kid followed slowly right behind. When the two goats reached another ledge only a short distance from where we first spotted them, both bedded down again with barely a glance at the predators. The wolves stared at the goats for several more moments before retreating down the mountainside and out of view. Seldom have I seen any response to dire danger as cool and nonchalant as that.

For the record, all of the goats we first spotted on that mountain beside Tuaton Lake were still present and accounted for when we broke camp a couple of days later. Their descendants are probably still living there, easily outwitting the wolves and anything else that threatens, except possibly the avalanches and terrible cold of winter.

With keen vision and great confidence in their nimble feet, mountain goats often move rather nonchalantly away from stalking predators, such as wolves.

8.
PRONGHORN

Wind Cave National Park, South Dakota, August 1980. Peggy and I had spent a week photographing animals in the beautiful Black Hills. We had watched magpies and bison, mule deer and prairie dogs, all at close range in our camera viewfinders. On the prairie dogs, especially, we'd shot much more film than we intended to. There are few inactive moments in a typical prairie dog town.

But one Wind Cave creature had proven elusive. We did not have a single exposure of pronghorns, even though they were quite plentiful and we saw them every day. Several even circulated around the prairie dog town, but these kept well beyond the range of our longest telephoto lenses. At first the situation was only frustrating, but eventually getting close to those pronghorns became a gnawing challenge.

One morning we spotted four animals—three does followed by a medium-size buck—and sat in our pickup watching them for a few minutes. Suddenly I remembered an old trick that Indians and early white explorers across the Great Plains had used to get near enough to shoot antelope with their short-range rifles. They would crouch low in tall sagebrush and hold up a strip of white cloth, letting it flutter in the wind. Because they are curious creatures, pronghorns would then come near and be converted to venison.

"What can we lose?" Peggy ventured.

Instead of cloth, I used a sheet of tissue paper and knotted this onto the top of a bush. Then we backed away about 100 feet and again sat in the pickup, watching.

At first the pronghorns paid no attention and we were about to give up our vigil as a wasted effort. That's when one of the females seemed to spot the tissue paper. Immediately she began to walk—stalk, it seemed, with stiff legs and head held higher than normal—toward our teaser. The other animals followed in single file directly behind, stopping for a second or two only once or twice. They walked directly up to the tissue, where all but the buck circled and sniffed at it. The trick worked so well that we sat for a few moments without lifting our cameras.

What followed was an interesting photo session. Once the prong-

The fact that pronghorn fawns lie motionless for a few days after birth enables biologists to locate them for tagging with the aid of trained dogs.

PAGE 162
This pronghorn buck energetically marked his Montana territory before finally settling down with some of that territory still hanging from his horns.

It is a late summer daybreak on the Wyoming high plains. A buck pronghorn stands up in its bed, stretches, and in the first light of day begins its wandering and feeding.

horns had ventured so near to our car, they no longer seemed nervous about our presence. We photographed them browsing, scratching, sniffing, studying the horizon, and even bedding down. They came even closer to the pickup before eventually wandering away, as calm and unhurried as wild animals can be.

But the truth is that we had not rediscovered a miracle strategy for attracting pronghorns. We tried the trick several times after that one successful encounter, and it never worked again.

Range and Number

The graceful, swift pronghorn, *Antilocapra americana*, is strictly a North American species, without counterpart anywhere else in the world. (Although many people refer to the pronghorn as an "antelope," it isn't really an antelope. Biologists call it pronghorn. So we'll use that name here.)

This tan-and-white native once ranged over an area of two million square miles in the south central part of the continent, from the tablelands of Sonora, Mexico, northward into prairie Canada. In the times of its greatest abundance, the pronghorn lived as far east as Minnesota and Iowa, and westward to eastern Oregon, Washington, and even into northeastern California.

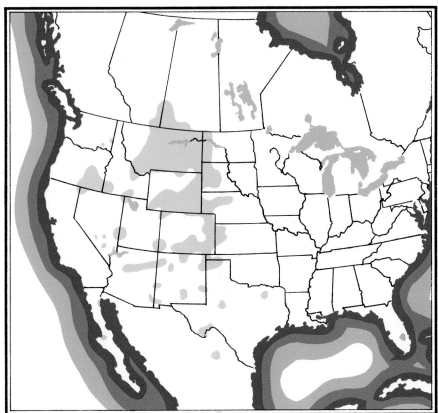

ABOVE
At a distance, herds of pronghorn may blend so well into their backgrounds that it is difficult to see them without a binocular or telephoto lens. Only the white rumps betray their presence.

MAP
Like many other mammal species, the pronghorn once ranged far more widely than it does today. From many millions in the 1800s, the pronghorn population dropped near 15,000 by the 1920s. Through restocking and other conservation moves, the population is now up to about half a million.

SHOT BY THE WAGONLOAD.   In frontier times, pronghorn must have been a lot less wary and even more curious than they are today. They were shot by the wagonload and the train-car load. They were then hauled into new boom towns such as Denver. There they were sold three or four carcasses for "two bits," a quarter then being the smallest coin in common usage in the Rockies. Colorado was still a territory in the 1860s and 70s and, like other states and territories, had no laws to prohibit hunting. Pronghorns were doubly easy to kill in excessive numbers during severe winters when savage storms swept the plains and trapped them in a few areas.

In the 1920s, it seemed as if the pronghorn would run no more across North America. Naturalist Ernest Thompson Seton estimated there were 40 million pronghorns on the plains in 1800. Whether or not that figure was accurate, the animals were astonishingly abundant; the diaries of John Charles Frémont and other early explorers note that pronghorns were as numerous as bison. One traveler in Nebraska said pronghorns were in such vast herds that "the foothills in the distance seemed to be moving." But the plundering we like to call winning the West changed all that in a few decades. Those same moving foothills were soon devoid of any animals at all except jackrabbits and coyotes. According to the first survey-census of the species ever done, from 1922 to 24, only 22,000 pronghorns were left in 16 western states. Some conservationists of the period believed the number was really closer to just 15,000. In Colorado, the peak population of two million fell to less than 1,000 by 1918.

NATIONAL REFUGES ESTABLISHED.   By 1925 the American public, in the first faint flickering of environment consciousness, began to realize that the plight of pronghorns was desperate and that Americans were on the verge of losing all of these animals forever. People realized that belated state laws protecting pronghorns were failing because they were not adequately enforced, and because such wide-ranging animals disregard state lines. Under pressure from conservation groups, a national game range and two national refuges totalling 800,000 acres were established on public lands, where the last remaining herds existed.

The Charles Sheldon Range, in the high plateau country of Nevada and Oregon, has more than a half-million acres of excellent wintering range. The Sheldon Refuge, adjacent to the Range, and the Hart Mountain Refuge in Oregon include favored fawning grounds. In winter most of the pronghorn in this area stay on the Sheldon Range, while in summer they roam widely, many of the does moving into the two national refuges to bear their fawns. Scarcity of water in the summer high country makes the Hart Mountain Refuge with its permanent water supplies attractive to the animals.

Shortly after these areas were established for the pronghorn, the Federal Government made funds available to Nevada and Oregon for wildlife purposes. These Federal Aid for Wildlife funds enabled the western states to hire biologists to study the pronghorn habits, diseases, and enemies, as well as to restock favorable ranges.

NUMBERS TODAY. Today this national wildlife treasure seems relatively secure. Next to deer, pronghorns are the most numerous of our large native mammals. The total population is close to half a million, with the largest number in Wyoming and second largest in Montana. Censusing pronghorn can be done with pretty fair accuracy because they live in open country easily transected by aircraft. Pronghorn are thus more economically live-captured and relocated from overstocked to suitable understocked habitat. As a result, wildlife biologists are better able to study and manage these animals than other large quadrupeds.

The Physical Specimen

The name *pronghorn* derives from the short pronglike branch midway up the front of the horns on bucks. These horns are unique among North American ruminants because their outer black sheath is shed annually from permanent horn bone plugs. After the shedding, new sheaths take about four months to form and harden. Some females also have tiny horns, shorter than ear length. Horns on a mature buck will measure

The pronghorn is the only North American native with forked horns, the outer covering of which is shed after the rutting season. The large, dark eyes provide phenomenal vision all around, keener than any man's.

about 12 inches along the front curve, but record-book horns have exceeded 18 inches.

Aside from horn length, bucks are also distinguishable from does by the black mask over most of the face and the black cheek patches on each side. Also, when irritated, approached by a rival, or when posing, a buck will erect the longer hairs of his mane. This posing is a peculiarity of the species, especially of males. Even fawns pose.

Although pronghorns are strikingly handsome, particularly against the skyline early or late in the day, their movements when walking make them appear almost goatlike. But the pronghorn, above all other big-game animals, was born to run. And run. Then the goatlike resemblance vanishes, and the pronghorn suddenly becomes an agile and graceful creature of speed—often too fast and elusive to follow with a camera mounted on a tripod.

Even the oldest, largest buck is a small animal, small enough at least to surprise most people seeing one at close range for the first time. A full-grown male will not exceed 125 pounds or 3 feet tall at the shoulder, and they average less than that. One Montana buck weighed what might be a record of 160 pounds on the hoof. Maximum longevity is nine to ten years, though few pronghorn live to half that age. The black horn sheaths are shed early every winter and grow back again by midsummer, a little longer than the year before.

Any pronghorn is marvelously adapted to racing swiftly across wide open spaces. The pronghorn can inhale great amounts of oxygen into relatively large lungs because of a trachea that is twice the size of a man's. A pronghorn's heart is also unusually large.

We often read of antelope being clocked at such preposterous speeds as 50 or 60 mph. Forty mph is much closer to the maximum speed

A quartet of pronghorn does stand motionless, studying the landscape ahead. A coyote may be crouched just beyond the next ridge. Pronghorn are eternally alert.

achieved. Perhaps more remarkable than the animal's top speed and how quickly it is achieved from a standing start, is the fact that a pronghorn can cruise along at 25 to 30 mph for half a mile or more. In addition, the pronghorn's large deeply cleft front hoofs, when widely spread, enable it to run swiftly over rocks, crevices, and parched ground without danger of tripping or falling.

Look especially at a pronghorn's eyes, which give it the keenness of sight often regarded as important to survival as its speed. Set back far on each side of the head, permitting the animal to scan horizon to horizon without turning its head, the eyes are as big as those of a saddle horse. It takes any predator, including man, an immense amount of effort, patience, and stealth to avoid detection.

A pronghorn's flashing of its white rump patch may be the most distinctive physical characteristic. When an animal is suddenly frightened, or, perhaps, merely senses danger, the long white hairs of the rump flare into a rosette that is easy to see from far away. Apparently this serves as a warning signal to all other pronghorn within sight. I have often spotted pronghorn in the distance only when one of the animals in a herd flashed and caught my attention. On the other hand, I have also seen those animals, when obviously frightened, run away without flashing at all.

Habitat

I've observed some other interesting and little-mentioned facts about pronghorns, at least about those in western Wyoming. In areas where the high plains begin gradually to tilt upward, toward the Wind River Mountains for example, many pronghorns also move temporarily into the foothills and even into quaking aspen country. Several of the largest buck pronghorns I have ever seen were on the edge of timber. In fact, I have photographed bucks bedded in secluded foothill areas more suitable for mule deer.

Another method of locating pronghorns, especially during dry seasons in dry areas, is to find any of the few and often hidden water sources. These include seeps and springs, small ranch impoundments, even intermittent streams in which a pool or two of stagnant water may still remain. Any pronghorn in the region will inevitably come to the water, however limited or impure.

Once during a dry fall, following a nearly rainless summer, Peggy and I found a seep that was not much more than a moist spot on an open slope in the Wind River foothills. No trees or shrubs marked the spot, as is often the case. Instead the ground all around was trampled bare by numerous visitors, which included mule deer and coyotes as well as pronghorns. One warm day, we watched the spot from a distance, and that was enough to convince us we had to move much closer. The seep had a steady stream of wild visitors.

In our pickup we have always carried a sheet of camouflage cloth among our assorted photo gear. Using the cloth, old parachute cord, rocks, and three discarded, rusting metal fenceposts that I found not far away, we built a crude photographic blind. We situated it within easy

telephoto-lens range of the seep, which we deepened—with our bare hands. Then we sat back to wait.

The first day was somewhat disappointing, because even thirsty pronghorn were too suspicious to approach the strange structure they found. But thirst won out, and by late afternoon one doe walked nervously but directly to the water and sipped while we ran film through two cameras. That broke the ice. Antelope and one deer came to drink at intervals until dark. We knew the next day in that blind would be a great one and planned to be inside it well before daylight.

Unfortunately, a violent storm struck during the night, with wind that destroyed our blind. We never saw our camouflage cloth again. Worst of all, the heavy rain that fell allowed all pronghorn in the vicinity to find water almost anywhere.

Fawns instinctively remain motionless in sagebrush cover for a brief period after birth. Soon they are able to follow their mothers everywhere and in only days can outrun the fastest man.

Feeding

Pronghorn have often been accused of eating grass needed for livestock, especially in dry portions of the West where grasses are at a premium. But the problem almost always stems from too many livestock on the range rather than competition from pronghorns. The species eats little grass except in early spring when the animals crop the new green shoots. In fact, pronghorns aren't nearly as destructive to western ranges as are an equal number of domestic sheep. Rather, pronghorns are browsers, preferring shrubs such as saltbrush, bitterbrush, and particularly sage-

We do not often see pronghorn around water because pronghorn country tends to be dry. But we look for the animals near water-courses during dry summers. We photographed this doe near water's edge.

brush. Pronghorns depend most heavily on this nutritious sagebrush both in summer and winter. They also consume many large plants that are poisonous to domestic animals.

The stomach capacity of a pronghorn is only about half that of a domestic sheep of the same size. Since pronghorn rely on sagebrush and similar browse plants for a large part of the year, they do not need the larger stomach capacities required by sheep and cows, which are grass eaters. Without sagebrush or its equivalent, however, the small stomach capacity of pronghorn (and mule deer as well) can be a serious disadvantage. Pronghorns cannot possibly eat and digest enough dry range grass to maintain themselves during long periods of extreme cold.

Harem Gathering and Dueling

September is an ideal time for pronghorn watching anywhere in pronghorn country. The combination of cool, invigorating weather with deep-blue autumn skies and dry vegetation is hard to match after a hot, dusty summer. With a good binocular or spotting scope, it is possible to watch the animals behave naturally from very far away. Bucks begin to gather harems of several does toward the end of August, and from that point onward bucks are on a constant patrol to keep females from straying and other bucks from intruding.

I have seen bucks fight, at times vigorously, but always before the harem gathering begins. After that a threat and a chase seem sufficient to drive rivals out of the vicinity. However, some other observers have reported frequent sharp dueling among males throughout the rut. Probably there is more fighting where the ratio of males to females in a herd is high, and much less conflict where the number of mature bucks is low.

Birth and Development

Minutes after a fawn is born, often into a windswept world, it staggers to its feet and nurses briefly. Next it crouches down at its mother's feet where its brown coat practically evaporates into the neutral tones of sage, bunch grasses, or rabbit brush. So begins one of the shortest childhoods among mammals. Within three days—four at the most—the fastest human couldn't catch the fawn no matter how hard he tried. Before the fawn reaches a year old, it will be the fastest long-distance runner in North America, and one of the fastest on earth.

But what produces the phenomenal growth and development of fawns? Why can they run so fast, so soon? The main reason is super-rich mother's milk, which is richest in nutrients just after a doe gives birth. Pressure to survive is another reason. The earlier in life a fawn can outrun a wolf, or more often a coyote, the better its chances of reaching adulthood.

DOE AND FAWN.   I have accompanied Montana wildlife biologists on spring censuses, both to count survivors after a brutal winter in the eastern part of the state and also to tag newly born pronghorn calves.

Wrongly accused of competing with domestic livestock for grasses, pronghorn eat shrubs and sagebrush as well as many large plants that are poisonous to livestock.

Counting is usually much easier than actually finding the many calves. One technician in the crew told me that he had seen a pronghorn doe strike with her hoofs at a golden eagle he had seen land nearby, driving it away. Thinking she was protecting a fawn, the technician hurried to the spot to tag it. He found the hiding fawn easily enough, but when it began to bleat pitifully in his hands the doe suddenly came after him. She stopped short of attacking and, instead, circled nearby snorting and hissing until the man went away. Most of the time does will stand by passively or even walk out of sight while a fawn is being marked.

FAWN-FINDING DOGS.   Jim Yoakum, a biologist with the federal Bureau of Land Management, may have been the first to use dogs to locate fawns during his research activities. Yoakum took advantage of the fact that, for a few days after birth, a pronghorn fawn will lie motionless and almost odorless until it acquires its running legs. He would venture afield with his black Labrador retriever almost exactly as a hunter strides through the country after pheasant or quail. With far better eyes and nose than its master, and its ability to cover much more terrain faster, the specially trained Lab would run up to a hidden fawn and either point to it or place its paw on the youngster's back. A well-trained "antelope dog" will not bite, mouth, or try to retrieve any fawn it finds.

Such encounters are probably stressful to the fawn, but not critically. Using a dog has enabled one biologist to catch and release 75 fawns in one month, or three times as many as he was ever able to find alone. One of the fawns was captured four different times, an indication that it was not injured either physically or psychologically.

What we have learned from this tagging may be partly responsible for the pronghorn's management and prosperity today. Only two of five subspecies—the Sonoran pronghorn of Mexico and southern Arizona, and the Peninsula pronghorn of Baja, California—are listed today as endangered or threatened.

Sagebrush and the Future

Like mule deer and sage grouse, pronghorns must have adequate, quality sagebrush habitat widely scattered in suitable locations if they are to continue in large numbers in the northern plains states. The problem of maintaining such habitat becomes increasingly critical each year as more and more sagebrush "wastelands"—as politicians and developers like to call them—are fenced, converted to farmlands and subdivisions, despoiled by mineral extraction, and covered by roads and reservoirs we do not need.

Pronghorns face many problems in late 20th-century America, but the continued sagebrush eradication, now taking place chemically, could be the ultimate threat to their survival. Sagebrush is important not only to pronghorns, but also to keeping vast areas from drying out too rapidly in summer. It offers shade and moisture beneath which grasses can grow. Sagebrush is also vital in preventing soil erosion by both wind and water. Yet the chemical spraying goes on, presumably to improve livestock grazing range, though we know that rarely if ever does sagebrush elimination improve grazing in the long run.

Sage grouse share range with pronghorn in the western United States. This male bird is strutting during courtship at a western Wyoming lek, or communal courting and breeding ground.

Highway and Track Deaths

Pronghorns also have trouble coping with traffic, both train and auto. One winter I followed a large truck and trailer rig barreling along over the speed limit southward toward Big Piney, Wyoming. Rounding a bend, the truck ran right through a herd of pronghorns standing on the snowy pavement, instantly killing six and badly crippling several others. I did not see the red taillight that would have indicated if the driver had even touched his brake. Every spring, melting snows reveal the winter-long carnage on many western highways.

On February 17, 1975, a Union Pacific locomotive may have set a dubious record for trains. Traveling at 70 mph west of Idaho Falls, near

Pronghorn are very swift afoot, although they need not always shift into high gear. Healthy animals can easily outdistance any predators. The greatest danger comes from hunters in pickups.

Good pronghorn country is almost always good coyote country. Coyotes exact a toll of fawns each spring, but the two species have evolved together over many centuries.

Craters of the Moon, Idaho, the train shot through a herd of pronghorns bedded on the track, killing 132 of them. But that was not all. Four cripples had to be shot, and when Game Department officials autopsied the dead they found that 25 of the does were carrying twin fawns—which increased the death toll to a stunning 186 animals in just a second or two. Recently I asked an Idaho conservation official about the incident, and he said that the herd in the area had not begun to recover from that and subsequent train accidents.

Having watched pronghorns for hours from a distance, I have concluded that they sometimes enjoy racing for the sheer exhilaration of it. On several occasions, I have seen pronghorns play a game of follow the leader, alternately walking and then running in single file wherever the leader decides to go. Another habit, and not necessarily a healthy one, is challenging passing cars, especially in unfenced areas. Typically the animals will bound alongside a vehicle, pass it, and then dart directly across the path of the car, only to stop on the other side and stare as it goes by.

Fences

Despite a pronghorn's phenomenal acceleration and speed, the species was not programmed to cross fences and certainly not to jump them. I have never seen a pronghorn leap or try to leap over even a low fence.

Instead pronghorns try to squeeze through or under it, which is not too difficult with the common three- or four-strand barbed-wire fences. Often they can go under while barely breaking stride.

Woven-wire fences meant to contain domestic sheep and goats are another obstacle altogether. They have caused the unnecessary deaths of thousands of pronghorns in recent years by preventing entire herds from migrating from summer to relatively snow-free winter ranges. A 9-mile woven-wire fence, known as Broadbent Fence, now blocks a migration route near Fort Bridger in southwest Wyoming, over which an estimated 2,000 pronghorns passed in the severe winter of 1983–84. Built over private as well as Bureau of Land Management land, the fence is not as high as an earlier fence in the Wyoming Red Rim country, which drew national attention. It is nonetheless a significant barrier that Wyoming Game and Fish Department biologists believe could be catastrophic. The biologists are currently seeking a solution to the fence, which is of the type that is illegal on BLM lands anyway.

The Toughest Predator

In the mid-19th century, the main predators of the pronghorn herds were the grizzly bears and wolves of the plains, plus a few coyotes (rarer then than now), bobcats, and golden eagles. All of those could only catch pronghorn fawns, or old, infirm animals. Against all those meat eaters, the North American pronghorn was well able to hold its own as it had for thousands of years since its evolution.

But survival is a lot tougher and a lot more uncertain, now that man has become the main predator and competitor.

9.
BISON

For weeks the news had spread across Comanche County, southwestern Oklahoma, that a train carrying a strange cargo would arrive in Cache, the drab and dusty county seat. So it is no wonder that on October 18, 1907, the station platform was a seething mass of ranchers and cowpokes, curious city people, and tribes of Indians. Among the latter was old Quannah Parker, last and best known of the Comanche chiefs. Late in the hot afternoon, the train finally arrived in town bearing a shipment of American bison (commonly called *buffalo*) in cattle cars. After five decades, the natives were finally returning home.

According to a St. Louis newspaper reporter covering the story, emotions were mixed, if not strained. Older Comanches openly wept at the sight. A few remembered the glorious days when bison roamed everywhere. But many ranchers from the surrounding Oklahoma hills did not feel any such nostalgia. They regarded bison only as competition to their own herds for the protein-rich grasses that grew to the horizons all around. Bison, like Indians, were something they could do without.

Altogether there were 13 animals—eight bulls and five cows—in the shipment. The bison were sprayed with crude oil to protect them from ticks, then released into a pasture of the newly established Wichita Mountains Wildlife Refuge. On that day they became the only plains bison to walk wild and free in the entire United States, except for 25 or 30 that still survived in Yellowstone National Park. We had come perilously close to wiping out all of the estimated 20 to 80 million bison that had inhabited North America when the first Europeans arrived.

Dr. William T. Hornaday, chief taxidermist of the U.S. National Museum, was the man responsible for the return of those bison to Oklahoma. Interested in bison and disturbed by their plunge toward extinction, Hornaday founded the American Bison Society. Later, as director of the New York Zoological Society, he urged President Theodore Roosevelt to create, by executive order, a large game sanctuary somewhere in the West to try to reestablish the species. In 1905 Roosevelt selected the site in the Wichita Mountains.

At that same time, Hornaday was studying the captive bison held in

A newborn bison calf is unsteady and nearly helpless for a brief period after its birth. It will not stray far from its mother, who will defend it with fury.

PAGE 178
Shown near one of the thermal areas of Yellowstone National Park, this bison covered with frozen mist conveys an almost ghostlike presence. Ironically, the saving of the last 30 bison in North America occurred at Yellowstone, allowing this bison to be more than an apparition.

the paddocks of the New York Zoo. He then picked out 13 of the healthiest animals, which had been captured originally from such widely separated ranges as the Texas Panhandle and the Manitoba prairies. All made the 1,800-mile trip to Oklahoma successfully, and later their numbers were supplemented with animals from other captive herds. But Hornaday's 13 became the foundation for today's population, which any visitors to the 50,000-acre Wichita Mountains National Wildlife Refuge can easily see—still living wild.

Origin of the American Bison

The tragic story of the American bison has been related so often that some people may not want to hear it again. But no wild creature has had a greater impact on life in early America, so we will cover the highlights here. The bison, or buffalo, *Bison bison*, apparently reached North America from Asia via a land bridge that once connected the two. Probably these ancestors of today's bison arrived during warm periods between the great glaciers that periodically covered the northern part of the continent.

    Through the centuries the bison slowly moved southward, retaining their heavy fur coats. When Europeans discovered America, the species ranged over a great portion of the continent. They had penetrated as far south as Mexico, maybe even to Guatemala, and as far east as what is now Pennsylvania, New York, Maryland, Georgia, and Florida. The wandering herds seemed to follow the river valleys and mountain

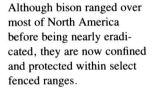

Although bison ranged over most of North America before being nearly eradicated, they are now confined and protected within select fenced ranges.

passes into the Pacific Northwest. But the species was concentrated in greatest numbers between the Rocky Mountains and the Mississippi River, from Great Slave Lake, Canada, southward to Texas. Perhaps no other part of the earth—not even Africa—ever produced a large game animal in such great numbers and density as the North American bison.

Throughout the summer, bison herds move constantly in search of new pasture. They do not hesitate to cross rivers and swim long distances if necessary.

Subspecies

The American bison is not related to the Asian water buffalo, nor to the African Cape buffalo. But it is related to the Eurasian wisent, which it closely resembles. There are two subspecies of American bison: the plains bison, by far the more widespread of the two, and the woods or forest bison, now restricted to a small range in northern Alberta and Northwest Territories, Canada. The two subspecies look so much alike (and readily interbreed) that it usually takes scientific analysis to tell them apart. But the skull and horns of the woods bison average larger, and a practiced eye might tell their darker coats from the lighter, perhaps sun-bleached coats of the plains animals. Woods bison also weigh more.

Importance to Indians

By the time Columbus arrived in the New World, the bison meant survival for many native Americans, particularly for the plains Indians. Few creatures have provided men with more. From hoof and hair to

horn, from brains to blood, no particle of a bison went unused. The species furnished shelter, clothing, tools, utensils, fuel, and medicines, not to mention the vast amounts of nutritious food. Even the entrails were eaten and relished. Bison parts were fashioned into ceremonial objects as well as toys for children. The chips provided fuel where little other fuel existed. Hide from a buffalo killed in winter became the warmest available robe or blanket, while the skin from a summer kill became the material for a tent. Hoofs served as hammers, horns as spoons, and ribs as sleds. Bison stomachs made water bags. And skins stretched over wooden frames served as small boats. Teeth were sewn as adornment into garments with bison sinew, while the tail could be a knife scabbard or a fly swatter.

So abundant were bison that no Indian conceived that the great herds might ever disappear. In fact, some tribes believed that bison sprang continually from a great hole in the earth, its location known only to the Great Spirit. But these Indians didn't account for the white man, who suddenly materialized among them.

Toward Extinction

Unlike Indians, Europeans exploited the bison for commercial gain and wasteful sport. That isn't to say that Indians always practiced wildlife conservation; when the pickings were easy, the plains Indians picked plenty. But there was something about the European slaughter of the bison that bordered on a crusade. In a little more than a century the uncontrolled killing reduced countless millions to a few strays. By about 1832, all were gone east of the Mississippi. Before 1900, the age of the American bison had ended everywhere. The species had been virtually eliminated specifically to bring about the final capitulation of plains Indians.

It is worth recounting some details here, in the hope that we never make the same tragic mistakes again. To begin, many bison were shot just to keep them away from lands newly plowed and planted by homesteaders. Military outposts and railroad building camps needed meat aplenty; but in time, the military killed the buffalo just to starve Indians and drive them to peace tables. The Union Pacific and the Atchison, Topeka, and Santa Fe railroads held excursions on which city people rode out to see just how many animals they could shoot. Summer vacations were spent shooting bison as a game, after which the carcasses were left to rot where they lay. Many European immigrants, suddenly free of restrictive game laws, used their brand-new guns to shoot animals (according to many accounts) apparently for the joy of seeing them stumble and fall.

Royalty—what might be called the 19th-century equivalent of today's jet set—came in droves to hunt bison and other game in wilderness America. Army generals entertained and encouraged them. Some hunting expeditions employed whole companies of U.S. cavalry to drive bison and cater to the whims of such guests as Grand Duke Alexis of Russia, all at taxpayer expenses. During his 1855 expedition along the Platte and Missouri rivers, for example, Sir George Gore of England slaughtered 2,000 buffalo, 1,600 deer and elk, 105 bears, and assorted pronghorn and wolves. Inundating the plains with that much carrion caused so many complaints that the U.S. Congress felt guilty enough to pass the first game law protecting the bison. But with no enforcement, the new law made no difference whatsoever.

The Hide Hunters

The so-called hide hunters have often been blamed for the extermination of the bison. These were mostly men who had been laid off by the railroads or mines and who desperately sought other income. True, the hide hunters took a toll in bison, perhaps a million animals a year for four decades. But their harvests were far removed from the wanton shootings just to watch large targets drop.

A typical hide-hunting team was composed of a gunner, two or three skinners, and a camp rustler. The gunner was the boss who calmly and accurately shot a day's quota without wasting ammo or time. The skinners removed the hides—which might have been worth two dollars apiece for the best—and loaded these onto wagons. The camp rustler cooked, pickled the tongues, smoked hams, piled and cared for the cured hides. Altogether it was hard, hot, time-consuming work for which remunerations were small.

Even after most of the bison had vanished, there were enough carcasses for homesteaders to make some money by bone-picking. Some lands were so littered with bison bones that these remains could be quickly gathered, hauled to terminals, and sold as cash crops. Whole trainloads of bones were then freighted east to be ground for fertilizer and used in bone china.

Savior of the Bison

At daybreak on March 12, 1894, a savage late-winter storm swirled over the high Yellowstone plateau of northwestern Wyoming. The temperature plunged to zero and a wild wind raged. U.S. army scout Felix Burgess debated whether to wait out the storm alone at the cold and empty Lake Hotel (the same one as in Yellowstone today), or continue on what seemed to be a hopeless mission. His left foot, from which the toes had been chopped by Crow Indians years before, hurt him terribly.

Burgess' mission was to protect the last herd of bison still clinging to existence in Yellowstone and, it was believed, the last free-ranging plains group surviving anywhere.

Using skis cabin-crafted of 12-foot lodgepole pine slivers, and a willow sapling as his single ski pole, Burgess headed eastward across the frozen outlet of Yellowstone Lake to follow Pelican Creek, also a solid sheet of ice. He had to keep traveling rapidly to keep from freezing. Several miles onward he stopped suddenly in disbelief; just ahead were fresh tracks of another man on skis. In that vast and lonely uninhabited

PAGE 184
Largest of the North American races of bison is the wood bison of northern Canada. A large herd of them survives in Wood Buffalo National Park in extreme northern Alberta.

BELOW
Winter is a severe test of every bison's vitality, especially on the northern high plains and in Yellowstone Park, where many bison congregate near the thermal areas, as shown.

wilderness, Burgess knew the trail could only have been made by one person—Ed Howell, the man he was hunting, the outlaw he was ordered to capture dead, but preferably alive. Burgess probably gulped before he turned to follow the trail, which was quickly filling with wind-driven snow.

Soon after Yellowstone became the world's first national park in 1872, the place was practically forgotten by Washington politicians. The remote park became a haven for poachers, especially for buffalo hunters who had already eliminated the immense herds that once moved across the land. Now the final remnant of those millions was also being slaughtered in Yellowstone, and in 1886 the U.S. First Cavalry was sent to the park to restore order, which they did.

Through vigorous patrolling and good administration, the soldiers soon eliminated all but a small amount of poaching along the park's boundaries, at least in summertime. But winter was another matter altogether. Horses were useless in a desolate barrens where more than 10 feet of snow sometimes accumulated on level ground, and subzero temperatures often persisted for weeks. The only possible way to patrol effectively was on crude skis, and that was very limiting. Even today with the finest modern equipment, cross-country skiers do not try to cross the Yellowstone Plateau without careful preparation and much apprehension. But apparently it was all in a day's work for Ed Howell.

If Howell had been anything except a poacher and an outlaw, he might today be regarded with Jim Bridger, John Colter, and Jedediah Smith as ranking among our greatest outdoorsmen and frontier heroes. Built like a bison himself, rough and tenacious almost beyond description, he was a master at survival. Earlier that winter, on makeshift skis he carved out with an axe, Howell had pulled a toboggan loaded with 180 pounds of supplies from Cooke City to Pelican Creek. That is about 40 miles as a raven flies, but more than twice that distance on foot over terrible terrain.

Howell didn't even own a pair of boots. Instead, he wrapped his feet in gunny sacking and then stepped into meal-sack bindings nailed to 12-foot skis. A cur dog was his only companion. He used an axe to make cut marks across his ski base underfoot to give a better grip climbing slopes. He also used resin to prevent snow from sticking.

Howell pitched a tepee at Astringent Creek (Pelican Meadow) and began shooting the bison that were wintering in the area. Only a few hundred animals were left, not only in the park but in the wild altogether. Shooting was good and often very easy. Howell would cache his skins and heads—in great demand by eastern taxidermists—deep in the lodgepole pine forests for later collection, beyond reach of roaming wolf packs. He never dreamed that anyone could find his isolated hunting camp, nor that anyone would try. It was the last and possibly the only mistake Howell ever made.

Traveling as fast as he could upwind in the storm, scout Felix Burgess followed the ski tracks to Howell's tepee where several green bison blankets were hanging. Pausing there, he heard six shots fired at intervals. Burgess moved off in that direction, skiing very slowly now, his heart pounding, but no longer from the exertion. Soon he came to an open white meadow stained red with blood. About 400 yards away, a man was bending down to skin one of five dead bison that littered the landscape. Howell's rifle was leaning against a carcass about 15 feet away, still within quick and easy reach.

Armed only with a .38 service revolver, Burgess had to make a perfect stalk or be killed. Fortunately a wind growing wilder by the minute blew in his favor and muffled his approach. So he crept unnoticed to within ten feet of the poacher, who was elbow deep in blood and gore.

"Drop the knife," Burgess said. To this day the knife may lie exactly where it fell nearly a century ago.

Burgess marched his prisoner first to Lake (an old summer hotel) for the night—he had skied over 18 miles since daybreak—and then northward toward Old Faithful, where he turned Howell over to a detachment of troops wintering in the area. According to Emerson Hough, a writer for the then-influential *Forest & Stream* magazine in New York, Howell was unrepentant and ate 28 pancakes for his first meal in captivity. Nothing is known of his fate thereafter.

Bison Today

Howell's capture just may have saved the buffalo in Yellowstone Park, and everywhere else, because about 30,000 are currently thriving and increasing in the wild. Winter visitors in Yellowstone regularly encounter them more than any other wildlife. But Burgess himself did not fare so well; the Army post surgeon had to amputate the badly frostbitten toes on the foot that the Crows had spared years before. Despite the important mission he accomplished, Felix Burgess' origin, and anything else about him remains a mystery. But all conservationists today are greatly in his debt.

Physique

Bison belong to the cattle family, which means they chew their cud and are cloven-hoofed. Both sexes have a single set of hollow curved horns. The immense bulls stand 5 to 6 feet at the shoulder and often weigh in at a ton or more. The most striking physical feature is the huge head and great hump covered with dark brown, woolly hair, which contrasts sharply with the small hips. Cows are somewhat smaller and less impressive.

SPEED AND AGILITY. A bison's great size and ponderous gait are deceptive, in that any of the animals can wheel about on a dime and quickly charge or break away. They have amazing speed, mobility, and agility. In deep snow, they can outdistance any other large mammal, except possibly moose, or a man on snowshoes, or a good cross-country skier. They can travel through powdery snow as easily as over dry ground. An adult bison is also surefooted enough to climb some of the same windswept mountain slopes where bighorn sheep spend winters. In fact I have found bison wintering along the same sheer cliffs with sheep along the Gros Ventre River in Grand Teton National Park.

Rivers pose little obstacle to bison, which are excellent swimmers.

ABLE SWIMMERS.   Bison are good and willing swimmers, able to cross any lake or river just to graze on the other side. They swim with only the nose, forehead, and hump above water. The tip of the tail, with its brush of hairs at the end, is also held high. Some of the most dramatic early bison accounts tell of entire herds crossing and practically filling wide and muddy western waterways.

SENSES.   A bison's sense of smell and hearing are better developed than its eyesight. When deprived of sound and scent by a steady wind at its back, a bison will often stand for long periods staring straight ahead, trying to catch any unnatural movement. Bison in parks and reserves are no longer wary of humans, but they should be regarded and approached only with extreme care.

Coat Shedding

Bison begin to shed their heavy winter coats in spring, and soon the hair drapes down all around them in tatters. The animals are least attractive at this time. To hasten the shedding and possibly to relieve an itching skin, they rub against boulders and trees or even roll on their backs on the ground. During the shedding process the bark is competely rubbed from many trees, killing them. In fact, the presence of buffalo can often be determined by the remains of trees thus killed.

Shedding of winter coats also renders buffalo more vulnerable to the hordes of biting insects that hatch in late spring or early summer. Insects can make many individuals more irritable than usual. To escape their tormentors, bison may spend a good bit of time wallowing in dust, sand, or muck, a common sight in many of the national wildlife refuges and parks where bison live. Some of the more frequently used wallows may be a foot or more deep and 15 feet in diameter.

Diet

Bison grass was and is a staple of wild herds. Two centuries ago it gave the plains from Mexico to Montana a parklike appearance. This kind of grass tolerates severe winters and long dry spells, as well as excessive tramping. But as the masses of bison disappeared, taller grasses won out

Shedding of hair in late spring gives all bison a shaggy appearance. A herd can be tracked by following the tufts of loose hair left hanging on low trees and brush in its wake.

Rolling and wallowing in dust is a frequent ritual of summertime. Wallowing may occur in established dusting places or anywhere the mood strikes and dry, loose soil is available. Rolling may dislodge ticks or other insects, or it may just feel good.

over buffalo grass, and vast tracts of it were lost. Today bison feed largely on grammas, bluegrass, bluestems, wheatgrass, and fescues.

It seems to me that bison invariably stop moving and feeding in stormy weather to stand or lie facing into the wind. When deep snow covers the range, the animals root deep into the drifts by swinging their heads from side to side to expose what is beneath. The muzzles often become encrusted with ice. Unlike other large mammals, bison do not use their hoofs to uncover grass.

The Herd Instinct

Bison definitely are gregarious creatures. Old accounts describe entire plains being black with thousands of animals in a single herd, and the passage of trains and covered-wagon caravans being held up for hours by passing "buffalo." During most of the year, however, the animals seem to separate into bull herds, composed mostly of older males, and cow herds, which may contain some young males as well as females and calves. Old warrior bulls often form still smaller groups, always isolated from the rest. It is difficult to say if they are tired and prefer the solitary life, or if they simply have been driven away by younger, stronger bulls.

Biologists now believe the great north-south migrations once ascribed to bison never really occurred. The animals did always wander a great deal in search of new grazing, or maybe to escape heat and in-

sects, but those movements could never be called regular migrations. The great bison herds, which were loose clusters of smaller herds, were followed by prairie wolves and coyotes and even by plains grizzlies, now long extinct. One biologist referred to these congregations as mobile ecosystems.

In winter, bison graze through deep snow by using their massive heads like bulldozers. Frost collects on the well-insulating pelage.

Mating

Late summer is a relatively good time for bison. Many of the troublesome insects are gone, the grass is tall and luxuriant, and the animals are growing fat on the bounty. It is also the time of the rut, when a wave of restlessness runs through every herd. Bulls that have remained separate and aloof from cows and calves during the rest of the year now circulate busily among them, searching for females coming into estrus. Normally very quiet, the males become quarrelsome and bellow hoarsely. Fights suddenly break out and end as quickly until from a distance the whole herd seems to be a seething mass. The annual rut, usually in August, is an interesting and often exciting spectacle to watch.

At the National Bison Range in Montana, I have seen lovesick bison bulls behave in a preposterous manner, shoving and charging, snorting, and digging up sagebrush with their horns until their entire manes and faces were covered with it. I have even seen two huge old bulls dueling head to head in the center of Mission Creek, while the cow both were pursuing slipped away into the center of a large herd.

The rut is a time of furious activity in the bison herd as bulls vie for supremacy and often behave in a bizarre fashion. Many fights break out in a cloud of dust. Bulls snort, pant, and paw the earth in a crazy competition for breeding privileges.

Births

Buff-colored or reddish calves are dropped when winter blends into spring, and later farther north. The calf is born somewhat apart from the main herd, with only the faintest suggestion of the characteristic shoulder hump. Soon cows and calves rejoin the herd where all calves gambol together. They begin to graze earlier than other large mammals, but may continue to nurse occasionally for as long as a year.

Rarely, an albino calf is born into a herd. I have seen two of them. One did not survive its first summer for reasons unknown. But the other grew into a magnificent bull that lived for 26 years on the National Bison Range where it was called Big Medicine, and where I photographed it. Big Medicine now stands as a full body mount in the Montana Historical Museum, Helena.

Longevity

An American bison matures at seven or eight years and might live to be 30 years old. With advancing age, the gait becomes slower and more deliberate, the horns grow rough and show greater curvature, and the coat becomes increasingly hoary. One old male in Yellowstone Park was at times mistaken in the distance for a silvertip grizzly bear.

"The Last Hunt"

In 1926 a select few, very wealthy sportsmen in America received a 12¾-by-9⅝-inch color lithograph with wax seal, signed by one A. H.

Leonard. Surrounded by hunting and Indian scenes, the message read:

> An invitation to _____ To become
> one of the hunters in the Last Wild Buffalo Hunt
> to Be Held in the United States.
>
> This hunt will take place in November,
> 1926, on Antelope Island in the Great Salt Lake,
> America's Dead Sea. Antelope Island is 30,000
> acres of rugged mountains—a range of peaks
> humped up into the lake near civilization—and
> yet apart from it; an island seldom visited. Truly
> no more ideal spot could be found for this last
> wild buffalo hunt. The days you will spend here
> will be all too few and too short, each crowded
> to the brim with thrills, health and happiness.
>
> Among the sportsmen of America there are
> but a few to whom the thundering buffalo herd
> is but a memory. The western plains, where
> once roamed countless millions, are stripped
> bare of this once prolific and noble animal. The
> waterholes no more know the sucking of their
> feet in the mud, and the dust clouds of the wal-
> low are seen no more. Gone to the happy hunting
> grounds of the Indian are the buffalo herds of
> yesterday.
>
> Knowing that you will avail yourself of this
> last opportunity to hunt, as did the builders of
> this empire, it gives me greatest pleasure to ex-
> tend to you this invitation.

Within weeks, bison calves double their birth weight and continue to grow rapidly. Mothers and calves rejoin the herds to travel and graze on the new grass of spring.

Before the white man "won" the American West and nearly eliminated the great bison herds, a unique relationship existed between bison and prairie dogs. Prairie dogs established towns where the bison had overgrazed.

I have been unable to learn if the bison on Antelope Island were really a secret remnant herd belatedly discovered, or if they had been stocked there close to Salt Lake City just for the hunt. Nor could I find out how many sportsmen accepted the invitation, how much they paid to hunt, nor how many bison they shot. But I do know that anyone who visits Antelope Island today would be lucky to find anything except whitened, decaying bones.

The bottom line, fortunately, is that the bison, or "buffalo," has been saved and can be seen in many parks of the American West today.

10.
MUSKOXEN

It was almost noon when we cleared the cliff of Cape Vancouver on the western Alaska coast beyond the mouth of the Kuskokwim River. Pilot Fred Notti aimed the small Cessna floatplane westward across Etolin Strait, the channel that separates Nunivak Island from mainland Alaska. Nunivak, our destination, was only a dark spot in the haze across the blue expanse of the Bering Sea.

"We're lucky," Notti shouted above the drone of the engine. "The ocean is very seldom as calm and the sky as clear as it is today."

For almost a week that August of 1968 I had waited impatiently in Bethel, a remote and dreary settlement 100 miles inland on the Kuskokwim, for a break in the weather that would permit me to reach Nunivak. I had a very important reason for wanting to go. That big bleak island, 70 miles long by 40 miles wide, was the home of the only wild muskox herd in the United States. I had long been fascinated with accounts I'd read about muskoxen; now I wanted to see and focus on these wild polar "cattle" myself. But I was beset in Bethel by day after day of cold rain, high wind, and a low ceiling, and I had even begun to consider giving up the hunt.

But at last the break came. After a violent storm one night, the overcast vanished, and by mid-morning Notti and I—joined by wildlife biologist Jerry Hout, then assistant manager of Nunivak National Wildlife Refuge—were winging out across the vast, empty tundra that comprises the Yukon-Kuskokwim delta. The shadow cast by our plane on the half-land, half-water wilderness below stirred up clouds of whistling swans, brant, geese, and sandhill cranes, as well as smaller waterfowl.

Normally turbulent and angry, Etolin Strait seemed as smooth as a millpond that morning, and the flight across it was pure pleasure. Less than a half hour from Cape Vancouver, we reached Cape Manning at the northeast point of the island. There Notti banked the plane southward. He had explained that the irregular coast below us was where we could probably spot the most muskoxen and also where he had the best chance to land in wind-sheltered water.

Dropping down to about 300 feet and flying at the slowest speed above a stall, we spotted the first muskox at a place marked on our maps with the jaw-breaking name of Kanikyakstalikmiut. Between there and

In some parts of its northern Canadian range, a wandering muskox may meet willow ptarmigan. This bird is in late summer plumage. It will turn white by the Arctic fall.

PAGE 194
I photographed this fine muskoxen bull by going ashore on Nunivak Island, Alaska, and stalking inland. This was my first good look at the great animal.

the southernmost tip of the island, we counted 75 more muskoxen, mostly in small bands. But not once was it possible for us to land near enough to wade ashore and stalk them with telephoto lenses. To do that we needed to find the target within, say, two miles of where the pilot could put down the aircraft safely without puncturing his pontoons. Even then, we had no guarantee that the muskoxen would allow us to walk close enough for pictures. I wondered if I'd bitten off too much to chew.

Flying westward now from Cape Mendenhall, we found more scattered herds and more adequately sheltered bays, but never the two close together. Then suddenly we came low over the Binajoaksmiut River, and bedded down beside it were five of the long-haired animals we were seeking. Near them Jerry Hout spotted a lone bull and beyond him another herd of six. The pilot shouted to get ready to land.

Notti put the airplane into a steep bank and we slipped down, softly settling at the mouth of the Binajoaksmiut and scattering rafts of sea ducks in all directions. Minutes later we were on shore, exchanging hip-boots for hiking shoes. Then we cut across country toward the muskoxen we could not see from sea level.

Even with heavy photo gear in our backpacks, in ordinary going a 1-mile hike is a snap of about 30 or 40 minutes. But on the spongy tundra of Nunivak it was another story. I figure we could have covered 4 miles on hard ground in the time it took to finally come in sight of the first herd, which we spotted from atop a high bluff.

Treeless terrain punctuated with snow patches stretched to the horizon. Just below us, pale as aquamarine crystal, was a necklace of pools connected by rapids. The pools were choked with dog salmon coming onto Nunivak to spawn. A family of harlequin ducks flushed and flew swiftly upstream. Altogether it was a magnificent scene of Arctic wilderness that I will never forget. But we didn't take much time to stop and admire it.

Quickly I slipped out of the backpack straps, dug out camera and telephoto lens, and made several exposures even though the muskoxen were bedded about 200 yards away. That was much too far, but at least I would have something (I thought) to show for the effort, even if we never got any closer.

As it turned out, we didn't get closer. Some vagrant wind must have carried our scent to the animals because suddenly they were on their feet, splashing across the river, over a knoll, and out of sight. We also tried to stalk the lone bull and herd Jerry Hout had spotted from the plane, but failed there too. Disappointed, we trudged back toward the anchored plane.

Early Efforts to Save the Muskox

Muskoxen, like bison—which they superficially resemble—are natives of both Canada and Alaska. It is believed that the last ones surviving in Alaska, a herd of about 18 near Chandalar Lake, were killed by trappers there during the 1890s. Later, great sentiment developed to reestablish the muskox in The Great Land and in 1930 the U.S. Congress appropriated $40,000 for that purpose. What followed is typical of what hap-

pens when politicians, rather than qualified experts, become involved in conservation procedures.

THE SIX-YEAR JOURNEY.

Animals to restock Alaska could have been obtained from neighboring Canada, and Canadian wildlife officials were willing to make a fair deal. Instead, 34 animals—17 bulls and 17 cows, of a different subspecies than the North American muskox —were bought in Greenland to begin a journey that covered 10,000 miles and six years. The herd went first by steamer from Greenland to New York, where the animals had to be quarantined. The next lap was a slow trip via freight car to Seattle. Then came another steamship cruise to Seward, followed by another railroad freight-car ride to Fairbanks. After several years in semi-confinement near Fairbanks, where their handlers considered them an increasingly dangerous nuisance, the muskoxen were started on their way again, first by train to Nenana, then by steamship to Marshall, and then by towed barge down to the mouth of the Yukon River at Kolik and out across Bristol Bay to Saint Michael, Nunivak. That is where the globe-trotting herd was finally released on July 17, 1936.

It is only by the narrowest of margins that those muskoxen made it to Nunivak at all. The decrepit boat, Meteor, which was towing the barge, came close to splitting in half during the stormy crossing. Ac-

It is winter on Nunivak, and a small herd stands to face me. The animals are unpredictable; fortunately for me, these chose to run away.

cording to the account of a deckhand aboard the Meteor, crew members had to man pumps for 14 straight hours to reach the safety of land.

NUNIVAK MUSKOXEN TODAY.   The long trek proved to be worth the effort. Nunivak muskoxen numbers today are estimated at about 1,300. During the last decade or so, a small number have been shot on a drawing basis by big-game trophy hunters, and some have been trapped for return as possible domestic animals to University of Alaska farms near Fairbanks. The wool or *qiviut*, a Greenland Eskimo word, may be the most luxurious of all animal wool and is the basis of a struggling cottage industry. But even with the horns removed, these animals tend to be as truculent as ever in captivity, and they are not as easily handled and managed as domestic sheep.

From the nucleus on Nunivak Island other Alaska muskox herds have been established on Nelson Island, the Seward Peninsula, the Cape Lisbourne-Cape Thompson area, and also on the Arctic National Wildlife Refuge.

Number and Range

Scattered populations of muskoxen occur throughout much of the Arctic, many the result of restocking.

The muskox, *Ovibos moschatus*, is really circumpolar in distribution and cannot be considered endangered, although it is fairly difficult to reach and observe anywhere in the wild. The current population of eastern Greenland is judged to be about 15,000. Populations totaling about 10,000 are scattered across Arctic Canada, with greatest numbers in the Thelon Game Sanctuary, Northwest Territories, on Banks and Ellesmere islands. Small herds also roam on Svalbard (Spitzbergen) Island, Norway, and on Wrangel Island and the Tamyr Peninsula, U.S.S.R.

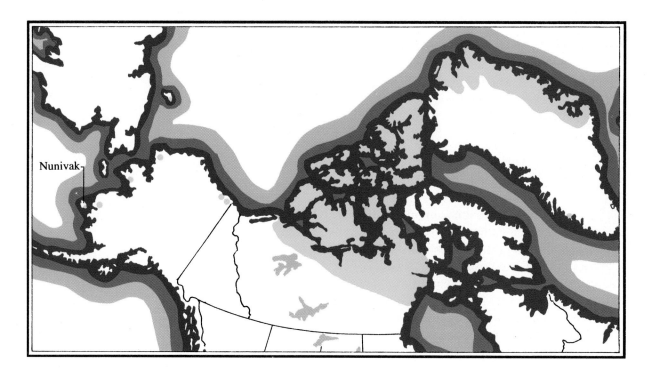

Nunivak

Physical Characteristics

Despite the name, the origin of which is a mystery, the muskox has no musk. The earliest Arctic explorers reported that there were musk glands in the skin, under the eyes, or in the feet, and even that the meat was tainted with the taste of musk. This myth has been repeated by adventure writers ever since. But the species is as odorless to human senses as animals can be. I have never eaten the meat, but a Nunivak Eskimo who has guided hunters and participated in live-capture operations assures me that the meat is "as sweet as the beef which is air-freighted from the Lower 48 onto the island."

A muskox's pelage is its most striking feature. Guard hairs as long as 2 feet, the longest of any North American mammal, hang down and almost touch the ground. They also form beards under the chins and, on males, manes that accentuate the humps over the foreshoulders. Once I tried to photograph a muskox cow standing still, and I was unable to see the nursing calf that stood completely hidden underneath the mother's long guard hairs.

Beneath the coarse, dark outer hair is the much finer and lighter-colored underhair or qiviut. From May through mid-August, muskoxen may have a ragged or mangy look because the old qiviut is shed during this period and is replaced by new growth. I've seen summer animals with long streamers of qiviut blowing in the Arctic wind until it eventually caught on vegetation or was blown away. I've also found Arctic bird nests that were lined with shed muskox wool.

It is easy to overestimate the weight of muskoxen due to their heavy pelage and long guard hair. Because of the logistics and handling problems involved, not many muskoxen have been weighed alive. But adult animals weigh only about one-half to three-fifths of what a bison of the same age and sex will weigh. A large male might reach 850 or 900 pounds and stand 5 feet at the shoulder.

Both sexes have horns. But after one year of age, sex-related differences are readily apparent. The horns of females are shorter and more slender than those of males. Male horns are shaped like a deeply dropping handlebar moustache, and in larger bulls they may exceed two feet in length, spreading out and then sharply downward from a massive base that protects the skull.

Feeding Habits

During the short far-northern summer, muskoxen prefer the moist habitat and lusher vegetation along streams. They move often in search of new green growth, nibbling the leaves of dwarf birch and Arctic willow as they pass. Game biologists have also noticed that these animals show a predilection for areas around coastal cliffs where the great colonies of seabirds nest each year. These same birds fertilize all vegetation adjacent to the cliffs, probably making it especially nutritious for any creatures that eat it. Some muskoxen are so eager to browse this heavily fertilized forage that they have fallen from cliffs trying to get at it.

Defensive Behavior

My unsuccessful photo hunt on Nunivak Island did dispel an ongoing myth about muskoxen. I'd read time and again how the animals always stand in a defensive circle when threatened—adults on the perimeter, calves safe inside—a tactic that many claimed led to their slaughter and eradication in many places. It was simply too easy to walk within rifle range and shoot the animals in their defensive posture. But now I know that explanation just isn't true.

The circle was indeed the ancient and effective defense against wolves, the traditional predator of the species. It was also the defense muskoxen used when pursued by Eskimos hunting with sled dogs and spears. But it obviously didn't work against men hunting with firearms and, more recently, on snowmobiles. What we might call "modern" muskoxen usually run immediately from the sight, sound, or scent of man.

Note the word *usually*. Not all male muskoxen choose to run. Biologists such as Jerry Hout who have worked with them for many years

Muskoxen know no natural predators in their Arctic range, except for the wolves that capture calves as well as very old and weak individuals. A band of healthy muskoxen are more than a pack of wolves can handle.

know that they can stand their ground, be troublesome, and even aggressive. Wolves, more than a few sled dogs, and their Eskimo owners have been suddenly charged and trampled to death by muskoxen.

The Hot Arctic Summer

Strange as it may seem, summers in the Arctic can often be uncomfortably hot, especially when the wind is still. Temperatures above 80° F are common for days and even weeks at a stretch. But these hot conditions do not seem to bother muskoxen; they may be pestered more by insects and parasites. A Nunivak native told me that he has seen muskoxen stand motionless in shallow, cold streams for extended periods when the day is extremely hot. But strangely, very seldom are the animals observed drinking.

Mating

The muskox rut on Nunivak begins in late July and continues through September. Bulls spend at least half of their time during this period in either sexual or antagonistic pursuits. Harems are acquired, and tremendous amounts of energy are expended to keep cows from wandering away and rival bulls from violating the harem. All the while, each slavering harem bull must continually test each of his cows for signs of estrus. I have never seen the muskox rut, but observers who have assure me that there are few dull moments.

According to biologist T. E. Smith of the University of Alaska, the clashes between harem bulls and their challengers are especially violent and similar to those of wild sheep. After preliminary bluffing, shoving, and posturing over an acre or so of level ground, the two bulls charge each other at speeds of at least 25 miles per hour, smashing together squarely on the horn bases. Smith has seen as many as 20 such impacts to settle a single fight. The only injury suffered most of the time is the loser's loss of dignity or ranking as it skulks away. Most of the time the original harem bulls, which are healthy 6- to 10-year-olds, emerge the winners.

The Long Hard Winter

Life is a lot harder in winter when snow, too deep even for such powerful Arctic beasts, covers much of the land. Herds are most concentrated where the snow is shallow and the crust easily fractured, or where savage winds blow away the snow altogether, exposing the grass, sedges, lichens, and reindeer moss that are the animals' main winter foods. On Nunivak, beach rye grass of the coastal sand dunes is the most important forage, but almost all exposed plants will be closely cropped by winter's end. An overpopulation or overconcentration of muskoxen can cause serious damage to the ecology of Arctic shorelines, as it unfortunately already has on Nunivak.

These muskoxen were photographed on an Alberta game farm where the males had become extremely truculent. In my opinion, this species should be left in the wild.

Muskoxen obviously need the super-insulation provided by their coats to survive the dark and frigid winters. But both native observers and biologists have noted that, especially in maritime areas, the extreme development of the pelage can also be a handicap and sometimes even a deathtrap. Balls of snow and ice collect on the guard hairs in sufficient size and weight to slow down the travels necessary to feed. United States Fish & Wildlife Service biologist, Cal Lensink, told me that muskoxen had become frozen to the ground and had perished by the rapid freeze-and-thaw conditions that occur on Nunivak Island.

Muskox Births

Single muskoxen calves are dropped sometime during April or May, when days are still cold and damp and when snow may still cover much of the lonely Arctic landscape. There are no known instances of twins, either in the wild or in captivity. For several weeks the calves remain very close to their mothers, often walking or standing on the flank that is protected from prevailing winds. At about one month of age, the young begin gamboling and playing with other calves, testing their legs and romping about in the middle of the herd. Females are not aggressive toward calves that aren't their own, as are mountain goats and some other wild mothers. In fact, life within an unmolested muskox herd during spring and summer is very peaceful.

Domestication

Throughout this book I have described the various environmental pressures and problems of North American horned and antlered animals. Loss of habitat, poaching, and destruction of the wilderness are a few examples. But lately muskoxen have suffered from still another of man's irrationalities—the obsession with live-capturing the animals in the wild to transform them into docile barnyard beasts.

Muskoxen are even less able than other larger mammals to withstand the stress of being chased far across the tundra by helicopters and snowmobiles, until they are driven half-dead into capture nets where they are drugged and dehorned. Perhaps half of those thus captured will live to be released on a farm, maybe one as far south as dry southeastern Idaho where the mercury in summer often rises above 100° F. All this for an "exciting" half-hour television program that does not show the dead animals. Or, perhaps, for a small amount of qiviut.

To me, the muskox is a remarkable, majestic wild animal that belongs living wild, exactly where it established itself. It does not belong, dehorned, in somebody's pasture or barnyard. If the herds need to be culled, as might well be the case, we can accomplish that far better by legal regulated hunting.

My Second Try

After failing to get any photos of the Nunivak muskoxen that summer day in 1968, we again climbed aboard the Cessna floatplane. After a short takeoff, pilot Fred Notti aimed the plane westward, following the island shoreline toward a place called Ikooksmiut. As before, we spotted a good many muskoxen but no place to put down near them. About that time I also noticed that a dark overcast was building up in the northern sky. Time for our expedition was running out. Then as he banked low over the sand dunes just beneath us, Notti pointed out the window.

"Lone bull," he shouted. I saw it standing beside a small stream in tall grass as we roared overhead. "Maybe we can land right there in the ocean," I shouted back.

Ocean landings in a light pontoon aircraft are rarely recommended because of the swells, which may be greater than a pilot can realize from above. But we made it down, bounced once, and taxied toward a somewhat sheltered bit of sandy shore. Luck had really smiled on us. A few minutes later Jerry and I were wading onto solid ground. I knew another year might pass without such an opportunity to land a tiny float plane on the open surface of the Bering Sea. Now we had only to scale the dunes to come within sight of the bull.

It was tough climbing in the fine dry sand, but when we reached the last ridge our quarry stood only 100 yards away, watching in our direction almost as if he expected us.

The bull gave me enough time to take several exposures before it bolted away. My heart sank because, again, that distance was much too far for any worthwhile photos, even with a 500mm telephoto lens. But

unaccountably the animal stopped running and turned once more to face us. This time, by taking our time and never walking directly forward, we were able to approach to within 40 or 45 yards of the muskox. There it posed for several minutes, turning from side to side before it ran away again. I now had my first—and still my best—muskox trophy on film.

Long guard hairs are underlain by finer, insulating *qiviut*. Sometimes the guard hairs become so weighted down with snow and ice that they hamper the muskox's movement.

A Game-Farm Bull

Later that summer, I visited a large game farm in western Canada where a herd of muskoxen was one of the feature attractions for visitors. The time was early September and the peak of the rut. As I watched the irritated harem bull prancing along a fence that separated him from another bull in the next pasture, a gamekeeper arrived with several bales of alfalfa to feed the animals. He said he was leery about going into the enclosure with the muskoxen—even more leery than when he had to feed the huge Kodiak bears kept in an enclosure nearby.

"Them bears," he noted, "are sweethearts by comparison."

The man wasn't joking, and in a moment I knew better what he meant. The harem bull had backed about 25 feet away from the fence and then launched a charge at the bull on the other side. He never connected, but he broke off a solid ten-inch post and changed the shape of the fencing. The bull then had to be tranquilized while the fence was repaired.

What a shame that fine bull wasn't still fighting and living free on Banks or Ellesmere Island, rather than being cooped up on a game farm surrounded by llamas and camels.

11.
EXOTICS

Early in 1984, an unusual shipment—five black rhinos—arrived from South Africa at New York's Kennedy International Airport. There the crated, tranquilized animals were transferred to an air freighter bound for Houston. Once in Texas the animals were divided up. A male and two females were trucked to a ranch in south Texas, while the remaining male and a female went to a ranch near Fort Worth.

The main reason given for shipping the huge and sometimes truculent beasts halfway around the world, from African to American bush country, was to save them. Black rhinos once roamed vast areas of eastern and southern Africa by the tens of thousands. Only a few hundred survive today. It is almost the end of the line for the magnificent species. Thus the new home in Texas does seem to be a way of saving them from African poaching for rhino horn and destruction of their native habitat. In Arab states around the Arabian Gulf, rhino horn sells for $9,000 a kilo. Elsewhere the animals have been shot and the brush cleared so that peanuts could be planted.

The relocation to Texas was the brainchild of American big-game hunters, ranchers, and zoo people for several reasons. But perhaps the main reason was that a lot of people, particularly Texans with the space and money to spare, simply like to see and raise rare, exotic wild animals. It is similar to collecting fine art or jewelry, only much more exciting. To me, no art ever created by man is more exquisite and more graceful than some of the wild animals now roaming free in Texas and elsewhere. In many cases, it may be a choice between survival on a Fort Worth ranch or disappearing forever.

Unfortunately three of those five black rhinos died of unknown causes. At this writing the others have not yet reproduced, which was expected. But for better or for worse—probably worse ecologically in the long run—a great number of exotics ranging in size from tiny Asian hog deer to white rhinos, the second largest land mammals on earth, are living and rapidly increasing in parts of wild North America. Exact figures are hard to come by, but many non-native horned and antlered species are now firmly established in the United States and Canada.

Sikas are attractive, extremely hardy deer of northeastern Asia that may be gaining a foothold in the North American wild. Experiments have shown that they are likely to threaten North American species through competition for food. In this case, whitetails are particularly vulnerable.

PAGE 206
The spiral-horned addax is an African desert animal that has been virtually wiped out in the Old World. But it has been given new ranges on Texas ranches.

The Boom in Exotics

It is difficult to say exactly when and how the boom in exotics was born. It may have been in 1929 when Robert Kleberg Sr. of the King Ranch bought eight blue bulls from a circus that had gone bankrupt in Corpus Christi. These are Indian antelope, also called nilghai, which are not handsome compared to other antelopes, and they are very prolific. Mexicans now working on the King Ranch, where the number of nilghai is about five or six thousand, call them valentinos.

It wasn't until the 1940s that a number of Texans began to collect the first herds of other exotics—now called Texotics in Texas—mostly on ranches in the central Texas Hill Country. Dick Friedrich probably was the pioneer, but in the 1950s Charles Schreiner III began to heavily stock his YO Ranch near Mountain Home with many different species, mostly for trophy hunting. Today Texotics are everywhere in the Hill Country, and it is not necessary to drive very far to see that these animals have often ravaged the countryside by overgrazing or overbrowsing it. They have also made it difficult for many native species of wildlife, especially whitetail deer, to prosper on the same ranges. (In *Erwin Bauer's Deer in Their World*, I describe scientific field experiments in Texas that revealed how whitetails were unable to cope with or compete with sika deer when the two were installed together in typical whitetail habitat.)

Numbers of Exotics

At intervals the Texas Department of Wildlife and Parks conducts a census of exotic mammals living in the state. In 1984, the figures for the three most popular species were 38,000 spotted (axis or chital) deer, 19,000 blackbuck, and 8,000 sika deer. But these figures are conservative. Most ranchers we've known who are seriously into Texotic man-

Barasingas, or swamp deer, are another species of India with a very bleak future. However, those on a few Texas ranches may prosper and multiply—and so sustain the species.

agement believe that two to three times those totals would be closer to fact.

We have seen an amazing variety of Asian and African animals now established in many areas of the American Southwest and as far north as Alberta, but mostly in Texas. These range from such rarities today as the red lechwe and roan antelope of southern Africa to markhors of the high Himalayas and vanishing barasingas of central India. One need no longer travel to Africa to see or even hunt such trophy species as greater kudus and sable antelope, aoudads, and Siberian ibex. That is now possible without leaving the American Noah's Ark.

A Texotic Ranch

One cool and crisp January morning, Peggy and I drove out across a south Texas ranch, and for awhile it was difficult to believe we were not on the Kalahari Desert of Namibia from which we had recently returned. A herd of gemsbok with long rapier horns grazed beside the rough and dusty track ahead. Suddenly two males slashed at each other in a dominance fight, faces flat to the ground. A herd of springbok watched the fighting gemsbok from a discreet distance. A little farther on, the Kalahari spell was broken as a herd of Asian axis deer stopped at our approach and studied us before moving away into golden sunlight.

Axis Deer

In all the world there is no more elegant deer and, perhaps, no more handsome horned or antlered animal than the axis deer, *Axis axis*, or chital. It isn't any wonder that from the time the Roman legions began to conquer the then-known world, axis deer were brought back to Rome along with other looted treasures, to live in game parks. The species has

My candidate for the most handsome, most exquisite of all deer is the chital, or axis deer. No wonder it is currently the most widespread and numerous of all exotics in America. The first herd established outside its native lands went to Hawaii.

PAGE 211

When the Far Eastern sika
deer are transplanted onto
American whitetail range, the
whitetail has trouble competing for food. Also sikas can
eat a greater variety of plants.

thrived ever since in European wildlife parks, where it is not too cold or snowy. Chital reached Hawaii from India in about 1867, as gifts to the King, and still live on Lanai, Oahu, and Molokai.

Chital retain their cream-spotted cinnamon coats all their lives. The belly, throat, and inside of the hind legs are white, which adds to the striking overall appearance. A chital stag stands about 3 feet at the shoulder and is 6 feet from nose to tip of tail. At about 200 pounds, it averages heavier than a whitetail.

Spotted deer are more herd animals than either the American mule deer or whitetail. I have seen them in groups of 40 to 50 in Texas and in India. During very dry seasons, herds tend to become larger as the animals travel to reach water, no matter what the obstacle. They are capable swimmers and in Asia will readily swim to escape pursuit by the tigers and leopards that prey upon them.

North American chital may have plenty of competition from both native and other exotic animals, but they hardly ever encounter a predator larger than a coyote, which might capture an occasional very young fawn. Even those losses are hardly noticeable since the species—or at least some animals in a herd—is capable of breeding all year long. I have seen bucks in various stages of velvet and with bare polished antlers during every season of the year. Velvet bucks, no matter how large, give breeding bucks a wide berth until the tables are turned in a few months. Of all North American antlered animals, a small race of whitetails in the Florida Keys are the only others that have no single breeding season.

Sika Deer

Sikas are deer of the Far East, where 13 similar subspecies are native from Tibet to Okinawa and from Viet Nam to Manchuria. Although many collectors and zoos in North America maintain pure subspecies of sikas, most of those now roaming America are some mixture of the 13 races. Generally speaking they are dark, slightly smaller than whitetails, and spotted, although the spots in some subspecies may be faint and even indistinguishable. And they are handsome.

We have seen most of the male horned and antlered animals fight, both during the rut and for dominance among a group and over food. But we have never seen any fight more savagely, more noisily, nor longer than a pair of evenly matched Shansi sika bucks in Canada. No joust rules were honored in a battle that may have lasted five minutes, yet surprisingly neither combatant was seriously injured.

Peggy and I have talked to several biologists who have worked both with sikas and native deer. Nearly all agree that, more than some other exotics, the sika could become a serious threat to northern whitetails and possibly mule deer if its range began to extend northward beyond Texas. Sikas can survive a series of bitter winters and deep snow, which other exotics cannot. They can also eat a greater variety of vegetation than whitetails and gradually starve them out. That may be true in the four Maryland Eastern Shore counties of Dorchester, Somerset, Wicomico, and Worcester, where released sikas are now very numerous.

Fallow Deer

Time has virtually obliterated the origin of the fallow deer, *Dama dama*, which probably was a native species of the Mediterranean or Middle East areas where it no longer exists in the wild. But because they are attractive deer on European hunting preserves and royal estates, fallow deer have more than just managed to survive. They now exist in at least semi-captivity around the world. Fallows are the most widespread of all exotic deer now living wild in the U.S. I have seen them in Land Between the Lakes, Kentucky, in Point Reyes National Seashore, California, and in Maryland, as well as very widely across Texas.

The so-called typical fallow is fawn-colored with white spots in summer, becoming gray-brown or even mouse-colored in winter. But no other wild mammal in the world has more color variations than this species. There are some herds of all-white and all-black fallows. There are also brown-spotted fallows, pintos, and even cream and silver-blue animals, which have resulted from selective breeding for centuries in European game parks and collections.

Given good nutrition and the freedom of an open range, a mature fallow buck is an impressive, wary animal with massive antlers for its modest body size. The antler conformation falls somewhere between those of a mule deer and a small moose, the antlers palmated at the tops. It is interesting that a deer that becomes so tame in formal deer parks can quickly become as shy and elusive as a whitetail when released into the wild. We have taken our share of fallow deer pictures, but the true trophy bucks have mostly eluded us.

The fallow rut occurs in late October or November in North America, at which time the males are very vocal. We have heard their deep, rolling grunts from at least a quarter-mile away. At first, we had to follow the strange sounds to the rutting bucks before we were sure what amimal had made them.

Full-grown sambar bucks may weigh as little as 135 pounds in the Philippines and as much as 600 pounds in India. Sambars are closely related to the American elk.

Sambar Deer

The sambar deer of India, Sri Lanka, and southeastern Asia is a semi-aquatic species that has also found its way to America. Similar in size to the American elk, the sambar deer may never become as popular as other members of the deer family because they seldom grow heavy or widespread antlers such as those of an elk of the same age or weight. Sambars also tend to be nocturnal, preferring life in damp and humid forests, as in one area of northwestern Florida where they may be becoming established.

During the rut, a sambar bull will select a territory and then spend a lot of time there bellowing. This sound may be less a challenge to other bulls, as is often written, than an invitation to cows. While some kinds of deer seem to do better in American surroundings than on their original home ranges, sambar deer do not, and—for the sake of native American wildlife—perhaps that is for the best.

No living person has ever seen Père David's deer, originally of China, living in the wild. These are on an Alberta game farm. Animals that survived the 19th century on an English estate and in the United States will soon be reintroduced into a Chinese national park.

Père David's Deer

Another species with a preference for soggy places, now showing up in North America, Père David's deer has been gone from the wild for over two centuries. Its very existence was unknown until about 1865, when a French naturalist-missionary, Abbé Armand David, happened to spy over the wall of the imperial hunting park in Peking, China, a sanctuary then forbidden to all outsiders. Père David noted that the deer grazing inside were different from any species known to science. Somehow he contrived to have a few live specimens shipped to zoological gardens in Europe. The species has since been known as Père David's deer, after the missionary.

In 1894, China was suffering from one of its chronic famines when sudden devastating rains flooded the Hun Ho River and breached the walls of the Peking hunting park. Hungry peasants slaughtered all but a few of the Père David's deer as they escaped. The last one in China died in 1921 in the Peking zoo. But those deer that had been shipped back to Europe by Père David prospered, and in 1964, four were shipped back home to China for a rare second chance at survival. I have also seen them surviving in at least three places in the New World: on an Alberta game farm, a Texas ranch, and a wildlife experiment station in Virginia.

Red Deer

Red deer, which once ranged from Scotland across western Europe and Russia to mountainous Iran, Tibet, and Kashmir, must be included on any list of Texotics. The species exists today on many private ranches, too often on the same ranches where American wapiti also have been stocked. It is difficult to tell the two apart. The result is that the two readily interbreed, and many of the animals now regarded as either red deer or elk are instead a mixture of the two.

A swift and shy antelope once abundant on Indian plains, the blackbuck is today more numerous in Texas. The black-and-white coat denotes a mature male.

Blackbuck Antelope

One reason often given to justify the proliferation of Texotics is that the surplus can be shipped to restock lands from which the animals were wiped out. Recently addax from Texas ranches have been sent back to Israel, Jordan, and elsewhere in the Middle East. Blackbuck antelope have been resupplied to Pakistan where they were all but wiped out. Almost surely more blackbucks live today in the Lone Star State than in all of India, where herds of them inhabited many open plains less than a century ago.

The agile and attractive blackbuck antelope owes its presence and popularity in North America to the fact that it is a fine game animal. A male reaches full maturity when its coat has darkened from the brown-and-white of females and young bucks to black-and-white. By then it is a wary target, difficult to approach. It is also very fleet afoot. The long spiral horns are coveted for mounting.

During one winter's photo safari, Peggy and I spent many days on a Texas ranch where blackbucks had been established for over 25 years and where they had seldom been hunted. We could always watch and study them from a distance, say from 75 to 100 yards, but that is not close enough when using a telephoto lens instead of a rifle. The black-bucks would even shy away from our blinds and, unlike the other exotics on the ranch, would not be tempted by a bait of grain. Only the whitetail deer were less accommodating. To tell the truth, trophy hunting black-bucks is a considerable challenge.

Desert Antelope

Two large desert antelope, the addax and the scimitar-horned oryx, have fared poorly in the stony, arid country just south of the Sahara in north-central Africa. Too much wanton shooting for too long, by desert tribes-

The scimitar-horned oryx from north-central Africa is another species that would be facing extinction had it not been transplanted in the American Southwest.

men and soldiers with automatic weapons, has just about finished them. But in Texas and scattered ranches of New Mexico, it is an entirely different picture. Here in a new home, both species are thriving.

Of the two, the scimitar-horned oryx is easily the most impressive. The long backward-curving horns may be almost as long as the animal's body, or more than 4 feet. Among our most vivid memories of camera trophy hunting was of a herd of scimitar-horns passing us one evening in silhouette, single file, against a scarlet south Texas sunset. It was the same scene a sub-Sahara wanderer might have seen a century ago, before the age of machine guns and all-terrain vehicles.

Gemsbok and Roan

The gemsbok, a close South African cousin of the scimitar-horned oryx, has increased enough from original stockings in New Mexico to justify an open hunting season. We have been able to watch and photograph in Texas the same roan antelopes that are just barely clinging to existence in eastern and southern Africa where they originated. In fact, the story of most exotics we find in North America today is a familiar one; survival depends on translocation to the opposite side of the globe.

Wild Sheep and Goats

Many kinds of the world's sheep and goats have joined other antlered and horned animals on the trip to the United States. Most are doing well, possibly too well. It doesn't take too many of these anywhere to reduce fragile habitat to bare or eroded ground. But they are in America now, and Americans had better make the most of them.

Barbary Sheep

The aoudad or Barbary sheep was among the first of the Texotics to be widely distributed, and its New World range now extends farther west into the deep, remote, and dry canyon country of New Mexico. There the terrain is similar to the aoudad's original habitat in the most remote mountains of the Sahara. This preference for the most hostile and inaccessible places in northern Africa has probably saved Barbary sheep from military vehicles, rapid-fire guns, and the same fate as the addax and scimitar-horned oryx.

Aoudads are sandy to brown in color and the soft hair is particularly long and apronlike on the chests of the rams. The males reach weights to 250 pounds or so and grow outward, slightly backward-curving horns that are massive enough to make valuable wall trophies. Our first impression was that these animals were sluggish and slow-moving, until we watched a herd of them traveling across sheer, crumbling cliffs in Palo Duro Canyon State Park, in the Texas panhandle. The truth is that all are nimble, good jumpers—at times jumping rather than walking across a gully or chasm in their path—and surefooted on the most difficult terrain.

Many wild animals blend in well with their environments because of a natural camouflage, but Barbary sheep are especially hard to spot on the wall of a Texas canyon where their coats blend into their backgrounds and the stark desert shadow. Beside the ability to stay unseen, an aoudad's hearing and eyesight are unusually keen.

Probably the first big game successfully introduced into the American landscape was the nilghai or blue bull, a native of Indian brush country now prospering in Texas.

Mouflon Sheep

Mouflon, wild sheep originally of Sardinia, Cyprus, and the Corsican mountains, are widespread and increasing in number on too many private lands in the Southwest. Males of these Mediterranean sheep are extremely elegant animals with the heavy curving horns that are characteristic of all the world's wild sheep. But the original strains have been so diluted with domestic stock for so many years that the species may not really exist any longer. Too many mouflon on the land can be devastating to the natural vegetation and cause serious erosion. In fact, mouflon sheep may be the most destructive of all the exotic mammals running loose in America.

Siberian and Persian Markhors

A number of the world's true wild goats have now been naturalized in North America, chiefly in Texas and New Mexico. Most widespread are the Siberian and the Persian ibex and the markhors. The latter is the largest and most sturdily built of all the wild goats, with the most spectacular horns. Males stand 3 to 4 feet at the shoulder and have long and rakish pointed spiraling horns that may reach 5 feet in length. A markhor's coat is long and silky. Billies have long beards that hang down over the chest. Here is a wild animal almost as magnificent as the precipitous land in highest Afghanistan and Pakistan from which it comes.

There were six races of markhors in Asia, but the origin of those in the United States today is unclear. Those in America may be of mixed race, having passed through zoos and other wildlife collections on the way to freedom.

Persian and Siberian Ibex

I have photographed the Persian ibex in its home in Iran, and in Texas. None of the horns on the billies we found in Texas could match some of the sweeping, laterally compressed scimitar-shape horns we noted some years ago on several splendid males in the mountains of northern Iran. This species has a reddish-brown coat in summertime, which turns

Corsican mouflon sheep now live in wild or semi-wild habitat in many parts of North America. Almost everywhere they reproduce so rapidly that they are destructive to the environment. The long, curving horns on this ram are of true trophy dimensions.

The markhor, a wild, high-altitude goat of central Asian mountain ranges, is living today on scattered Texas ranches—habitat totally unlike the native haunts.

grayish-brown in winter. Males have darker faces and beards. They are incredibly good climbers almost from the time they are born.

The species of Eurasian ibex, *Capra ibex*, has been subdivided into several races, and the largest of these is the Siberian ibex of outer Mongolia, eastern Siberia, and western China. Like the others, this is an animal of sheer, dizzying high places, both in Asia and in New Mexico where it now lives on public lands. The Siberian has the best chance of all the wild goats to establish a permanent home in America.

Future for America's Exotics

How far should we go in providing permanent homes for alien animals that have no home anywhere else? Admittedly it is pleasant and exciting to drive over a ranch in Texas, and feel you are in Africa. But the problem is that a lot of landowners who populate their properties with exotics have no concept of what this introduction can do to the quality of their land. As an extreme example, New Zealand had no native large mammals when it was first settled by Europeans. There, large animals from moose and mule deer to Himalayan tahr and chamois were introduced wholesale, without any thought to consequences. The animals nearly destroyed those two beautiful island environments before a massive campaign had to be launched to eradicate nearly all of them.

Only time will tell whether the introduction of exotics into wild America was wise or not.

12.

WITH A CAMERA

This chapter discusses an absorbing activity—trophy hunting with a camera, which can all too easily develop into a passion. If you read this chapter, you may find yourself buying cameras and equipment you never dreamed of owning. You may also find yourself stalking magnificent creatures in heat and cold, winter and summer, in some of the most beautiful places in the world.

Man—The Trophy Hunter

Man has always hunted animal trophies. More than 20 centuries ago hunters were scratching records of their trophies on cave walls in what is now Poland and France. The early Mogul emperors of India and the Samurai of Japan hired the best artists of the times to depict scenes of trophy hunts for gaur and tigers and great stags. Recently on a remote canyon wall along the Yampa River, Colorado, Peggy and I found an Anasazi hunter's depiction of a desert bighorn sheep with massive curving horns.

Trophy hunting survives today wherever there are big-game species to stalk. That isn't surprising because the wilderness of North America still contains huntable populations of these mammals. Some hunters have a powerful drive to take animals with the largest horns or antlers; but as compelling as gun or archery hunting can be, trophy hunting with a camera is far more challenging and often far more addictive.

Even the oldest black-powder rifle with open iron sights, or an old hunting bow for that matter, has a longer effective range than the telephoto camera lenses that can be carried into the field. Thus a photogra-

Many people find trophy hunting with a camera more challenging and addictive than hunting with a firearm. Here a bull moose dines unaware of my presence.

PAGE 220
Conrad Rowe cleverly combines use of a camera riflestock of his own design with a detachable monopod for added stability. Peggy and I instead prefer a tripod.

pher must approach much nearer to his target. Such conditions as rain and snow, deep shade and lack of sunlight, are not nearly as troublesome to a rifleman as to a photographer. Nor does a rifleman have to carry a tripod or waste time focusing and setting exposures.

Advantages of Camera Hunting

Camera hunting also has its advantages. For one thing, it can be a lot cheaper than rifle hunting. There are no licenses to buy, no permits to be drawn, and no outfitters to be hired. Nowadays a top outfitter in America's top game areas sells only very expensive hunts. Camera trophy hunters are not restricted by open seasons or bag limits, either. They can keep shooting long after firearms hunters hang up their equipment for another year. Photographers can film rare and endangered species as well as the common ones. Even after shooting a trophy or two, they can continue to seek a still better one.

Summed up, trophy hunting with a camera and telephoto lens is as thrilling, as demanding, as rewarding, and as competitive as any sport can be. I have been a big-game hunter both ways since the early 1930s (although most recently with camera alone), but none of the mounted trophies can match the satisfaction I got shooting some of the photos in this book. I strongly recommend trophy hunting with a camera for outdoor people of all ages who prefer to remain forever young and active.

The basics and many of the techniques of good wildlife photography are covered in detail in my two earlier books in this series, *Erwin Bauer's Deer in Their World* and *Erwin Bauer's Bear in Their World*. In this book I will concentrate especially on how to camera-hunt trophies of the larger species. I will also describe some of the newest photo equipment available, and how you, as a trophy hunter, might make the most of that equipment.

Two Ways to Camera Hunt

There are two ways to go trophy hunting with a camera. The first, which offers the poorest odds for success, is to proceed as a meat or gun hunter would by hunting just anywhere the game exists. This becomes a combination of potluck and dependence on your hunting skill. Any big game living in any area where it must survive an open hunting season every fall is likely to be far too wild to be easily photographed by a roving photographer. Of course you may get a few pictures of fleeting targets far away, but unless you genuinely relish challenge, this is usually not the way to go.

There is one exception: Some of the photos in this book are of completely wild, hunted whitetail deer and moose, which Peggy and I photographed from blinds, often using bait. But more about that kind of photography later.

A second, far better way, is to photograph trophies only in sanctuaries, parks, and reserves where firearms shooting is not permitted or perhaps where hunting pressure is light. Even on many refuges, some of

While some large mammals can be easily photographed near the roadside, you must make plenty of footprints and go climbing after others. Here I'm after goats in the Beartooth Mountains.

the animals may remain quite shy. But almost always in such places, a cautious photographer has a better chance of making a closer approach. As most large mammals become gradually more accustomed to people, they become less likely to bolt upon seeing people. A list of all the public places where it is possible to hunt North American big game with a camera follows this chapter. I am anxious to hear from readers about any other good spots, public or private.

Choosing the Right Location

Shooting a trophy picture—such as a shot in which a huge bull elk stands perfectly focused in bright sunlight, bugling in an exquisite mountain setting—might well be a product of luck. But so-called "luck" is really the "residue of design." After all, locating the right place to set up a tripod requires a knowledge of ethology (animal behaviors) much more than it requires luck.

Knowing Your Subject

Ethology is the study of animal movement, activity, and general behavior. It allows you to predict, with reasonable certainty, what a creature will do next. Knowing where a herd of bighorn sheep will be or on what trails caribou migrate at a given time is the major requirement for successfully photographing them.

Although Peggy and I understand the motivations and behaviors of native big-game animals, we are often surprised by individuals that do not follow the normal pattern. So above all, we always approach all of our targets with the caution and respect they deserve. Virtually all wild creatures have what might be called a flight-or-fight range. When a photographer approaches that boundary, the animal will either charge or run, and nearly always the latter. Neither makes for an attractive photo. To successfully shoot trophy pictures, we rely on our not-so-secret weapons—a telephoto lens, mounted to a sturdy and compact 35mm single-lens-reflex camera (35SLR).

The 35SLR

A skilled woodsman and ethologist can, in time, take acceptable wildlife photos with any kind of camera. Pioneers of this sport once used cameras so heavy and so cumbersome that I can only marvel at their photographs today. But without question the finest camera for shooting wildlife today is the 35SLR, of which a bewildering number of brands and models are being marketed. The 35SLR is the best choice because most are compact, sturdy, reliable, fast to operate, and relatively lightweight. Our 35SLR cameras feel comfortable in our hands, which is important. They function as well in very cold and damp weather as in the hottest, dustiest situations.

Newer and better 35SLRs are being introduced so frequently that it is impossible to keep up with developments. Many are so virtually foolproof and fully automatic that a photographer merely loads film, aims, and shoots. Today's models even focus automatically. But we have not yet succumbed to any of the automatic devices except the built-in exposure meter. This probably stems from a reluctance to relinquish control. It could also be the stubbornness of age—of becoming set in one's ways.

A 35SLR (single-lens-reflex) camera employs one lens for previewing and then exposing the film. Many other cameras manufactured today use two separate lenses for these functions. With a 35SLR, the same picture you see in the viewfinder is projected onto the film. The mirror reflects the image into the viewfinder while you compose your picture and then flips up out of the light path when you release the shutter to expose the film.

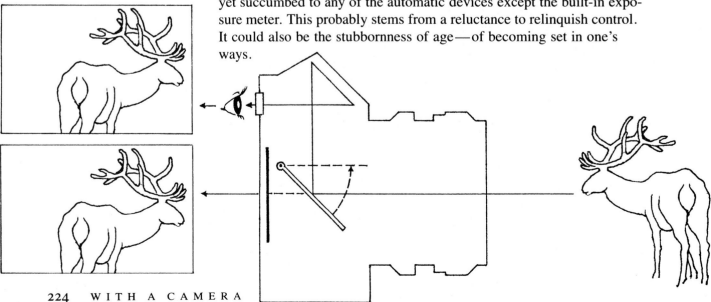

Peggy and I always have readily available three or four replacement 35SLR bodies that are exactly the same. These cameras have viewfinders in which exposure information is flashed around the edges of the picture. That way we can take notice of shutter speed and lens opening while we focus and compose a picture. The feature is particularly convenient when light conditions fluctuate or when the subject moves.

The large lenses in the back row are all Nikons; left to right they are 600mm f/4, 80–200mm zoom f/3.5, 400mm f/3.5, 200–400mm f/4, 300mm f/2.8. The small lenses and accessories in front, left to right, include the following: Nikkor 55mm f/3.5, Vivitar 5600 flash, Nikkor 35–70mm f/3.5, Vivitar 70–210mm f/3.5. The two small lenslike accessories at the extreme right are Vivitar 1.4× and 2× tele-extenders.

Motor Drives

All our cameras are equipped with motor drives, which offer many advantages. Following any action, such as a prancing bull elk, is smoother with a motor drive because it leaves you free to concentrate through the viewfinder on focus and composition. The distraction of manually advancing the film with a thumb is eliminated. Motor drives can also rewind film, but the heavy drain on the batteries is too much to pay for the slight saving in time over rewinding by hand.

The disadvantages lie mostly in the extra weight, the extra mechanism that might fail at a critical time, and in the metallic clinging noise that spooks some animals the first time or so they hear it. But that first cling has often made an animal look up and seem more alert for the second cling an instant later.

Lenses

Because wildlife photography is something I do almost daily year-round, I now carry and occasionally use what might be regarded as a gross assortment of lenses. I will list them here just for the record. All are, of course, interchangeable among Peggy's cameras and mine. The lenses: A 55mm f/3.5 (macro); 35–70mm f/3.5 zoom; 70–210mm f/3.5 zoom; 80–200mm f/2.8 zoom; 300mm f/2.8; 400mm f/3.5; 200–400mm f/3.5 zoom; and a 600mm f/4.0. On a 35SLR camera, a 200mm lens gives 4× magnification, a 400mm gives 8×, and a 600mm telephoto gives 12× magnification.

Here I'm equipped for a day of trophy hunting far afield. My basic equipment includes a 35SLR, motor drive, long telephoto, and tripod with a ball head. Extra lenses, film, and snacks are in the multi-compartmented backpack.

TELEPHOTO LENSES. There are so many good telephoto lenses on the market that selection for wildlife photography can be bewildering. The only point on which all veteran photographers agree is that the lenses need to be sharp, lightweight, and fast. By fast, I mean that the lens gathers a lot of light quickly, allowing faster shutter speeds than slower lenses do under the same light conditions. Basically, telephotos are lenses with longer-than-normal focal length (the distance from front lens to film). They magnify the size of the subject on the film. So, with a telephoto, from great distance you can achieve a subject image size as large as if you were up close to an animal with a normal 50mm lens.

TELECONVERTERS. A teleconverter or tele-extender tucked in a pocket or backpack can be worth its weight in gold in wildlife photography. I realize that not all photographers will agree that a 1.4× extender, which instantly converts a 300mm into a 420mm lens or a 400mm into a 560mm, is worth the price. Even fewer believe that a 2× (which doubles the focal length of any lens) is worth carrying, mostly because it penalizes the photographer two f/stops when installed. But we have employed both of them—particularly the 1.4×—so many times when a target was far away that we never travel far without them. Of course there are sharp and not-so-sharp teleconverters for sale. We use Vivitars and find even the 2× satisfactory.

ZOOM LENSES. Our 35–70mm zoom is the lens we use when we can approach very near to very trusting subjects, or when the background or scene is more important to us than the animal itself. Day in and day out, however, we use the 80–200mm zoom most often; it's our workhorse. We may hand-hold it; support it with a monopod; or rest it on a rock, tree limb, or car window sill. The 300mm is the best optic I have ever aimed at trophy game, but the 300mm and the 400mm remain more and more in their waterproof, dustproof cases. We now depend on the newer 200–400mm, almost always mounted on a tripod. It is an extremely fine lens that gives a trophy hunter flexibility in composition, but at 6½ pounds it is too much bulk to carry far slung over the shoulder. The 600mm is heavy insurance, not practical to carry far into the field but handy for reaching out to fairly static subjects that won't let you get close.

A tele-extender, or teleconverter, inserted between the camera body and lens increases the focal length of the lens, thereby enlarging the subject. The 1.4× tele-extender shown here increases the image size by 40 percent. A 2× extender would double the image size.

The 120–600mm zoom recently introduced by Vivitar deserves some special mention here. It is compact at only 14 inches long, weighs only 32 ounces, is sharp at all focal lengths, and is inexpensive at about $700. It is also easy enough to use. The single drawback is its slow f/8 lens speed at 600mm. But for backpacking and filming in the mountains away from trailheads, this 120–600 may be all the glass a trophy hunter really needs on bright days. Fine with fast film, it even fits into the average daypack—a boon when climbing for goats or sheep.

Rifle-Stock Mounts

More than a few crack professional photographers use their 35SLR cameras mounted on rifle stocks, telestocks, or similar shoulder devices to stalk big game. There is no denying their good results, obtained by pulling a trigger rather than touching the shutter release. Some lightweight telestocks can be adjusted to the photographer's physique, but we have never found any that exactly suited our own requirements. Whenever possible, we use a tripod or, less often, a monopod.

Camera Supports

To be useful for a mobile and energetic photographer, a lightweight, tubular monopod should telescope from about 24 inches (suitable length to fit in a backpack) to about 60 inches. The best monopods are not too expensive and can double as a walking stick or wading staff to cross streams with swift water.

TRIPODS.   There is no substitute for a firm tripod that is easily and quickly set up when the trophy is far enough away to require a long telephoto. The main disadvantages of a tripod are its weight and bulkiness. A really sturdy tripod restricts your maneuverability and limits hiking distances. And I also shudder whenever I think about the number of great pictures we've missed while moving a tripod to a "better" spot.

Not many tripods offer the necessary compromise between supportive sturdiness and lightness of weight. Some of the so-called backpack tripods, which are ultralight, are sturdy enough with long lenses only if their legs are *not* extended. Thus, these lightweight tripods force you to sit or kneel, even when the wind is not blowing.

Most of the time we lean toward heavier tripods such as the Gitzo, because the legs on these can be extended individually and positioned so that the tripod can be set up quickly on uneven ground, even on the steepest of mountain slopes. That last feature is a very important consideration in big-game hunting. When buying a tripod—or any photographic equipment, for that matter—it is good advice to try out many different kinds for comparison. Take your camera and telephoto lens to a shop where many tripods are sold and experiment with them on the spot, satisfying yourself that the one you select will also be the best choice in the field.

Len Rue Jr. demonstrates left-handed, left-eyed use of the Rue riflestock camera mount. Most people would use the device right-handed. Riflestocks are especially useful in photographing running game and in quick shooting of game you encounter on the trail. I rank the Rue stock as the best.

Peggy Bauer demonstrates the use of a monopod, a handy and lightweight accessory that might be used in place of a tripod. Collapsible monopods fit easily into backpacks.

TOP LEFT
Here I'm using a Vivitar 283 flash unit mounted behind a Fresnel lens that focuses the flash beam, illuminating wildlife as far away as 50 to 60 feet. Also shown is a telestock. My teleflash system was designed by George Lepp and modified by Tom Wiancko.

TOP RIGHT
Although a sturdy tripod limits my maneuverability in many situations, it provides the essential support for longer telephoto lenses that would otherwise exaggerate the slightest of lens movement, causing fuzzy photos.

BOTTOM LEFT
As shown, a good tripod also allows you to sit or lie while taking photos.

BOTTOM RIGHT
A quick-release mounting feature on a tripod head can save fumbling around when you could be shooting exciting pictures.

TRIPOD HEADS. The head that connects the camera to the tripod is at least as important as the tripod itself. Most of the inexpensive heads now available will not do the job. A good head firmly holds a camera with lens pointed exactly where you want it, without slipping even slightly or requiring great force to screw it tight. A good head also allows you to shoot from an absolutely steady base while you smoothly follow the action. Ideally this following—panning—up or down, left or right, should be so fluid that you are able to concentrate entirely on composition and focusing.

MOUNTS. A quick-release camera mount is certainly preferable to a screw-in mount on a tripod head. With adapter plates mounted on all of our cameras and longer telephoto lenses, this quick release feature permits us to mount or change the camera on the tripod within seconds. Once in place the camera is unable to work loose as it might on a screw-in mount. The best tripod head we've found to date is the expensive Monoball. We are currently looking into professional video camera tripods and heads as an even better combination.

How to Handle a Camera

No matter how good the equipment, none of it is any better than the person using it. Practice, more than any lens that costs a month's pay, is what produces impressive pictures. Study the equipment manuals carefully and then practice shooting the camera with various lenses at home and outdoors at every opportunity. Practice following the kids, family pets, passing cars—anything, at first with an empty camera until the camera feels like a part of your hands so that you operate it almost automatically. Zoos, game parks, local nature reserves, rodeos, and even football games are good places to go for extra practice.

How to Get Sharp Focus

Concentrate especially on rapid handling and on fast, but sharp, focus. Probably sharpness should be your first consideration, because your speed comes gradually with handling. Since big-game encounters are usually fleeting, your degree of success often depends on how quickly you react.

Often all that prevents a fair-to-middling picture of trophy wildlife from being outstanding is lack of sharpness, or detail. It makes no difference whether the subject is a moose, goat, bison, or elk. Careless photographers usually discover the fuzziness only when the slide is projected onto a screen or is returned as an enlarged print that magnifies the "soft" focus. In this case, the culprit may be camera movement, or the shutter speed may have been so slow that it caught movement of the animal. But far too frequently an otherwise good picture fails because the photographer does not focus on the proper spot.

When shooting big game or any living subjects it is very important to concentrate and focus on the eyes. The next time you are composing a picture in the viewfinder, note the bright bead or catchlight in your subject's eye. This reflected highlight is what makes any creature look truly alive and alert. If you turn the lens-focusing ring until the highlight is pinpoint sharp, you will instantly transform an ordinary shot into a dramatic one. If the eye of an animal is in genuinely sharp focus, the rest of its face will also be in focus, or sharp enough to rivet any viewer's attention.

Focusing on an animal's eye is not always easy. Many animals, human as well as wild ones, are not comfortable under prolonged eye or camera-lens contact. The natural reaction is to look down or away. In addition, many animals avoid looking into the sun when possible. So to get highlight in the eye, a photographer must be very patient and have a few tricks of his own. For example he must wait for the split second when an animal looks up and brings its eyes out from deep shadow. Sometimes you can help by moving a short distance away to change the camera angle or by appearing disinterested. This often causes an animal to look up.

Catching the highlight in an animal's eye is easiest during early mornings and late afternoons when all outdoor photography is immeasurably better anyway. On mammals, we occasionally find it just as easy to focus on the wet nose or on the whiskers. But focusing on an eye, as a rifleman aims at the small bull's-eye rather than the whole target, is a discipline that always pays off in more exciting photos.

FOCUSING SCREENS.   There are other ways to achieve or improve sharpness. One means may be optical. You may have the wrong focusing screen inside your viewfinder, or one that is not compatible with your vision. Although all popular 35SLR camera systems offer a variety of interchangeable focusing screens, most cameras come equipped with a matte Fresnel field surrounding a circular split-image range-finder spot. Perhaps some photographers really do prefer these split-images out of habit. But our advice is to discard these abominations

Photos of animals are greatly enhanced if you focus on the eyes and manage to capture sunlit "catchlight."

Most cameras come equipped with a viewing screen that offers two or more optional viewing spots to help you focus, including a split-image bull's-eye surrounded by a matte Fresnel-type screen (top drawing). But I prefer the faster focusing of a plain, fine-ground matte field, often called a Type D screen (bottom drawing).

The longer the lens and the slower the shutter speed, the greater the need for rock-steadiness to avoid image blur. Natural and make-shift camera and lens supports, as shown above, can offer good stability when there's no time to set up a tripod.

immediately and substitute a focusing screen with an overall fine-ground matte field often called a Type D screen. The latter will make focusing quick and easy, rather than a challenge.

EYEGLASS PROBLEMS.   Wearing eyeglasses can make accurate focusing difficult too, but after you explain your problem, any good optometrist can prescribe glasses that better match your lifestyle. If you wear glasses, be sure to encircle the camera eyepiece with a rubber ring to protect against scratching your eyeglasses—a constant hazard with plastic lenses.

Most 35SLR systems also offer a selection of eyepiece correction lenses called diopters that screw into the viewfinder eyepiece. These permit both nearsighted and farsighted users to view and focus sharply without their glasses. A soft rubber eyecup that prevents extraneous light from entering the viewfinder can also be helpful. Or, try wearing a baseball-type cap with a long bill to reduce glare and, ultimately, eyestrain.

CAMERA MOVEMENT.   Something as basic as holding the 35SLR camera properly can reduce camera shift and, thus, out-of-focus pictures. Stand comfortably, feet apart. Cradle the camera and lens in the palm of the left hand, using the thumb and forefinger to focus. The right hand holds the right side of the camera body, the forefinger on the shutter release button. The shutter speed is set at 1/250th second or faster if the exposure permits. This way the weight of the camera system rests in both hands, allowing you to concentrate entirely on what you see in the viewfinder without shifting your hands.

Yet we rarely shoot by hand-holding the camera while standing up if it is possible to sit or to steady the camera against some stationary object. A rock, a tree limb, a fence post, or the window sill of a car (with engine turned off) are a few examples of easily available supports that help eliminate movement. When sitting down, you can use your drawn up knees and your body as the three legs of a tripod.

Conditioning for the Outdoors

Far too often, Peggy and I have spotted trophies standing way off at a lung-busting climb or hike away. Sometimes we've been able to reach the spot in enough time and with enough energy left to collect the trophy—sometimes not. But our average has been good. Thus, good physical conditioning year-round is another of the factors that are more important in taking professional quality wildlife photos than the brand of the camera.

EXERCISES.   Hiking and jogging are among the easiest ways to assure better physical condition. Peggy and I walk wherever and whenever it is at all feasible. Bicycling would also work. And we climb staircases rather than use elevators. A photographer friend of ours always mows his lawn on the run, both to stay in shape and to have more time left for prowling in the woods. Weight lifting and pushups are good

conditioners. When Peggy found some of our longer telephoto lenses too heavy to handle easily, she began working out with dumbbells.

PHOTOGRAPHING ON SKIS. Winter is a season when it becomes too easy to relax. But in the snow belt, winter offers a matchless opportunity to get the best of all exercises, as well as outstanding wildlife pictures, by cross-country skiing. Dollar for dollar, the best money we ever spent on outdoor equipment went for our first pairs of cross-country skis. With them we have been able to explore exquisite places where moose and elk winter—places inaccessible by any other means except showshoes.

Hiking and Camping Gear

Backpacks, beltpacks, camping gear, and especially footwear are items a serious trophy hunter must regard at least as important as his camera. Proper footwear is critical. Although it is possible to shoot many North American big-game animals through a car window without ever departing from paved roads, not many of the photos in this book were made without walking at least a short distance. Most were made after long hikes or climbs. You can't do that very long unless properly shod.

More than other outdoor photographers, a trophy hunter usually needs a tent and other camping gear, especially if he does not have a camper vehicle. The best opportunities to shoot are almost always during the first hours after daybreak and the final hours before dark. Camping makes it possible to be at a good site during these crucial times without having to make a long drive in darkness. We have often been able to camp almost within sight of the spot where bull elk bugle every September, and where desert whitetail deer come to drink at dawn.

BOOTS. Lately our wanderings in North America have been out of a VW camper van that is ideal. Always handy in the van's compartments

ABOVE LEFT
Prior physical conditioning pays off. It allows you to reach remote game areas with enough energy remaining to take pictures.

ABOVE RIGHT
Winter in some areas is the best time of all to go trophy hunting on snowshoes or cross-country skis. Peggy and I prefer the skis because they are lighter and let us cover more ground quicker. Cross-country skiing is an easy skill to learn.

are two pairs of sturdy, ankle-high Danner and Dunham leather boots with lug-type soles, well broken-in through daily use at home. We have tramped thousands of miles comfortably in our Danners, which are our basic three-season hiking and climbing shoes—that is, when the weather is dry and above freezing. We also carry two pairs of rubber hipboots and, for winter, two pairs of insulated, cushioned, waterproof Dunham Duraflex boots. The Dunhams keep our feet warm during sub-zero camera hunting. We may also wear cross-country skiing gaiters when the snow is more than boot-deep.

CAMERA VESTS.   The farther you leave roadside and vehicle behind, the more vital becomes a beltpack, backpack, vest, or some other device for toting gear. Forget about those camera bags, however chick and cleverly advertised, that sling over a shoulder. A multi-pocketed, compartmented vest with waterproof drop seat is a much more useful item if you're camera hunting in warm weather or feel restrained by a backpack. It puts everything from film and filters to snacks and mosquito repellent right at your fingertips. Veteran wildlife photographer Leonard Lee Rue of Blairstown, New Jersey, manufactures the best of these vests.

BELTPACKS AND BACKPACKS.   A beltpack may be a better choice for trophy hunters who want to cover a lot of ground quickly, and perhaps do some steep climbing, without carrying too much gear. Beltpacks fit snugly around the waist and keep a number of accessories and film always handy. But if I'm spending anywhere from a half day to several days in the field, my choice is an external-frame backpack. The variety of backpacks on the market is as bewildering as the number of cameras and lenses. Before buying a new one, tryout as many models as you can, always fully loaded with the equipment you know you will carry, plus a little extra for insurance.

Our own backpacks are complete with wide, padded hipbelts. Cinched snugly over your hips, these belts transfer nearly all of the pack's weight from high up on your shoulders down low to your hips—greatly reducing the effort, backstrain, and general fatigue of backpacking. For overnighters our packs contain heavy lenses, tripod, sleeping bag, tent, and food.

PACK CONTENTS.   Numerous compartments in our packs keep film and lenses separated from each other and all other items. We also include the following standard equipment: a poncho or ultralight rainsuit, high-energy snacks, insect repellent, a baseball hat, gloves, waterproof matches, compass, drinking cup, and, most recently, a hand-pump water filter and purifier. The latter is insurance against giardia, a parasite that is appearing in more and more once-pure water sources, even in remote wilderness.

CANOES AND RAFTS.   A cartop canoe or inflatable raft is another item that can give a trophy hunter an edge. We haven't used our own canoe nearly enough, mainly because there never seem to be enough days in the year, but the prospect is never out of our minds. We have

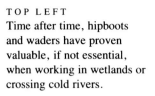

TOP LEFT
Time after time, hipboots and waders have proven valuable, if not essential, when working in wetlands or crossing cold rivers.

TOP RIGHT
The type of gaiters used by cross-country skiers is excellent over pants and boots when photographing in snow or wet vegetation.

MIDDLE LEFT
For many kinds of camera hunting, a ventilated vest like this one is the handiest possible equipment carrier. It has many pockets for lenses, film, and filters, a drop seat, shoulder straps, and a place to carry a tripod. The vest is designed and sold by mail order by Leonard Lee Rue III, the widely published wildlife photographer of Blairstown, New Jersey.

MIDDLE RIGHT
I'm equipped here for a day of spring photography in Yellowstone Park. Besides photo gear, the backpack contains food, a warm jacket, gloves, a wool cap, and a small water purification unit.

BOTTOM LEFT
A waterproof roomy carrier such as the Rec-Pac is handy for canoeing as well as portaging and short backpack trips, even for equipment storage in camp or vehicle.

BOTTOM RIGHT
Another ideal carrier for a modest amount of gear is a belt, or fanny, pack. This roomy pack is easily twisted from the rear, where it rides, to the front for access.

A canoe is an often over-looked means with which to photograph some wildlife, especially moose. It offers an easy, quiet approach.

used a canoe to quietly approach both moose and brown bears within range of a short-focal-length lens. But since we shoot more in mountains than in wetlands, such opportunities for us are relatively limited.

Blinds

A big-game trophy hunter has far less use for a blind than other nature photographers do, mostly because trophy hunting is an active, rather than a waiting, game. You seek your trophies by first finding them—perhaps with binoculars at very long range—and then by making tracks. Yet we've often used blinds, especially when we were after whitetail deer during hunting season. A few of the pictures in this book were shot from a portable ground blind, from a fixed tower blind (in thick brush country for whitetails), or simply from a portable ladder tree blind that we backpacked in.

The best portable tentlike blind is available from Leonard Lee Rue, Blairstown, New Jersey. The blind weighs just 9½ pounds. Instead, though, Peggy and I often use our camper van as a photo blind.

CHOOSING A PORTABLE BLIND. The ladder blind we use is made by Baker Tree Stands of Valdosta, Georgia. The best of the portable ground blinds is the Ultimate Blind sold by Leonard Lee Rue of Blairstown, New Jersey. It can be fully erected in two minutes, ready for use. Eureka! Tents of Binghamton, New York, also makes an excellent portable photographic blind.

Trophy Hunting Strategies

In North America, trophy hunting with a camera calls for two opposing strategies, depending on the place. To shoot in open hunting areas requires that you be as furtive and inconspicuous as possible as long as you are in the field. In other words, you must employ all the stealth, woodcraft, camouflage, and stalking skills of a good firearms hunter, and then some. The object is to approach within camera range of an animal before it sees you—and then get in as many "killing" shots as you can. As I mentioned earlier, this kind of hunting is a great challenge.

But if you score on a monarch bull elk or bighorn ram this way, you have a trophy truly worth bragging about.

SANCTUARIES.   Photographing in sanctuaries calls for an entirely different technique. In sanctuaries we *never* try to keep concealed from the target. Instead we strive to remain within our trophy's view—not so near that it moves away, but always so the animal can keep an eye on us and know what we are doing. Animals in sanctuaries, as elsewhere, can be exceedingly shy. Yet enough of them permit a photographer to approach within range of a telephoto lens, provided he stays in view. By staying in full view and never making any sudden, unnatural motions, we have been able to approach within amazingly close range.

There is a refuge in Montana where both whitetail and mule deer may be found. The does are usually not difficult to photograph, but the bucks are. Some photographers have never been able to shoot the bucks using conventional stalking techniques. But Peggy and I have gotten shots of several good bucks close enough and sharp enough for magazine covers, simply by sitting down, fully exposed on a mountain slope where the bucks fed at daybreak. Eventually they practically walked on top of us when climbing toward their daytime beds. Buck pronghorns anywhere are much less wary of photographers they can clearly watch. I believe it's a mistake to try stalking a pronghorn unseen, even a pronghorn that is used to seeing people.

I'm also convinced that virtually every trophy I ever want to shoot will, by superior sight, scent, or hearing, discover me at least as soon as I spot it. So I would only be kidding myself if I tried to approach unseen within the limited range of my longest telephoto lens. For this reason, we do not use or dress in camouflage cloth, though many successful wildlife photographers even wrap lenses and tripods with it.

Camping in the heart of big-game country is as pleasant as it is convenient. We work hardest early and late in the day. In the middle hours we relax and check over our equipment.

MALES MORE WARY?   In most wildlife sanctuaries, male animals are seldom more wary than females. Yet if it lives in a remote part of the park rather than in a locale frequented by people, one animal may be a lot more wary than another. Generally, the longer a sanctuary has been closed to hunting, the more confiding large mammals tend to be. During the fall rut, males are so preoccupied with breeding that they pay far less attention to photographers than during any other period. As you might guess, rutting time is ideal for photography. In autumn all males, whether they have reached full maturity or not, are more active, more visible, and more handsome. The weather tends to be bright in most places, and autumn colors enhance the backgrounds.

THE APPROACH.   Never advance purposefully and directly toward any buck, bull, or ram, even in a national park. Ninety-nine animals in a hundred will turn and drift away. That other one may come after you. Either way you get nothing on film. Instead, proceed on an angle, pausing often and closing the distance slowly almost as if you are planning to pass by the animal. We never stare or seem to concentrate on our target. In fact, we try to avoid eye contact. Patience is the greatest virtue in this kind of trophy hunting. If you take your time, you increase the chances that you will soon be standing close enough so your trophy fills much of your camera viewfinder.

SIZE OF HUNTING PARTIES.   With few exceptions—notably Yellowstone Park, where elk are accustomed to whole platoons of photographers nearly surrounding them—one person can approach most big-game animals better than two. By extension, two more easily than three, and so on. Many large mammals tend to be intimidated in direct ratio to how greatly they are outnumbered. A single photographer alone can move much nearer to most if not *all* deer than can two photographers approaching from opposite directions. Also, cutting off an animal's line of retreat—such as blocking a deer in an open meadow from dense forest, or a ram from open slopes to steep cliffs nearby—can cause otherwise calm animals to panic. The difference between success and failure—between a trophy shot of moose antlers and one showing a moose's rear end—is proper tactics.

Bad Weather

Bad weather occurs in all seasons, and every nature photographer is eventually condemned to inaction during a downpour. Some passing storm fronts offer unique chances to shoot pictures in eerie light, with briefly sunlit foregrounds against dark and threatening skies.

Particularly when the day is overcast, or when sunny periods are mixed with cloudy ones, we often sit down within sight of game and look away until the clouds pass. Many animals accept that and are quite calm around a sitting human they can watch. But dropping down on all fours for any reason is another matter, perhaps because it suggests some predator. It is one of the best ways I know to spook an otherwise confiding animal into running away. But when I'm just sitting, some animals may remain in the vicinity just to be able to keep an eye on me. More than once we have had bighorn rams move to beds nearby or to positions where they could watch us a little better.

But mostly we hope for good weather and try to schedule our travels to take advantage of it. That means keeping an eye on the sky for signs of immediate change, as well as checking long-range weather forecasts.

HOW TO PREDICT A STORM. To a great extent, you can predict local weather based on natural signs. High clouds are unlikely to dump rain or snow on you soon, no matter how ominous they appear. But beware of low, dark cloud banks. When leaves begin to show their undersides in a fresh wind, deer and elk may become more nervous and more active for a short time because a storm is in the offing. When the storm strikes, they stop moving about. Odors are so much stronger before a rain that you can sometimes smell it coming. So at that time, an animal's sense of scent is also keener than usual.

BEING PREPARED. Peggy and I have often been caught in bad weather far from shelter, and so we take careful precautions now. We always carry enough plastic bags (ziplocks, usually) in our belt or backpacks to protect camera gear and exposed film from rain. Just a little moisture seeping into moving parts of cameras or lenses can cause terrible problems. Whenever we do stay out to work in the rain, we work under tough, waterproof ponchos that will not puncture even when we press through brittle brush.

Sudden bad weather is always a possibility, especially at high elevations. We always carry all cameras, lenses, and film in waterproof plastic bags to keep them clean and dry.

A Trophy Animal

But what is a trophy animal—or trophy photograph—anyway? Like beauty, a trophy is in the eye of the beholder. Any photo that genuinely thrills the shooter is a trophy, no matter how modest the antlers or horns. But a serious trophy hunter, who will accept only an exceptional animal, may have to do a little extra study of the Boone & Crockett Club's *Records of North American Big Game* and visit a museum or a trophy collection to make measurements and comparisons.

In my experience, a trophy-class whitetail is the hardest to find and then photograph. A muskox or caribou trophy would be tougher only because of geography and distance. Whitetails, on the other hand, exist within a short drive of most people in North America. But for all antlered animals, it is necessary to see both a side and especially a front view to determine spread and mass. Let's use the whitetail as an example.

WHITETAIL DEER. An average buck is about 15 inches thick through the chest, and roughly the same distance between ear tips when the ears are cocked normally. If, from the front view, the antlers extend several inches beyond the body or ear tips on both sides, and multiple points are visible, you are looking at a good specimen at least. If some of the points visible from the front are a good deal longer than the ears, you are looking at a very good specimen, and it is time to start exposing film. Heavy dark antlers are a good sign. It's also promising if one

buck's antlers are obviously much heavier and more massive than those of a nearby buck, which himself also looks good. Focus on the bigger one.

BIGHORN SHEEP.   The horns on a trophy bighorn sheep should make almost a full curl, each almost as thick in cross section out near the "broomed" tips as at the base. Try to get a look at the ram from both front and side, as you would with antlered animals.

PRONGHORN.   A buck pronghorn is difficult to evaluate in the field before it bounds away. When viewed standing head-on, head in normal position, the vertical distance between a buck's muzzle and eyes is about eight inches. To be a trophy, the pronghorn should sport horns that are twice as long as the distance between muzzle and eyes—15 inches or better. Large forward-jutting prongs also contribute to trophy quality.

ELK.   When viewed from the side and walking, head up and muzzle forward, antlers of a trophy bull elk should extend rearward from one-half to two-thirds of the animal's back. Each antler should have at least six points. If you aim at a bull with seven points per side, you're assured of trophy quality.

CARIBOU.   A caribou's shovels over his face are usually an indication of overall antler mass. If it has two palmate shovels, each the size of a large man's arm and hand fully extended, odds are good you will be bragging about the trophy quality.

MOOSE. To be considered a super Alaska-Yukon moose, the area of each palmate antler must be at least equal to the size of the animal's face in profile. Of course that makes it necessary to look at the animal from both side and front, if possible when its head is lowered.

All the above guidelines are only the simplest rules of thumb for hasty use. Really, the only way to become an expert at assessing trophy animals in the field is to compare as many of them as possible consistently. Come to think of it, that task could become an extremely absorbing one.

Getting Within Range

BAITING. Most North American big-game species can be coaxed, enticed, or otherwise drawn into photographic range by baiting, calling, or other trickery. Though baiting is illegal in national parks and other public refuges, it is usually lawful to place bait for whitetail deer on private lands, with permission from the landowner. In fact, regularly placed bait of pellets, corn, or sweet potatoes over time remains the best way to attract a large whitetail buck within close camera range. On private lands, the same salt blocks meant for livestock sometimes attract moose or elk as well.

SUBTLE SOUNDS. We often use subtle sounds to gain the attention of an elk or deer. Loud noises such as foot-stomping, shouting, or whistling are likely to alarm an animal or, in some species, cause it to react aggressively. Instead, try closing your jacket zipper, scraping a fingernail over rough cloth, crushing a film box in your hand, or rubbing a lugged boot sole in the dirt. These noises usually cause an animal to be curious, at least long enough to bring its eyes into sunlight.

We have noticed that many wild animals look more alert immediately after hearing the click of a shutter and especially the noise of a motor-driven camera. Therefore we immediately shoot a second frame. This technique works amazingly well, particularly with animals that have not yet encountered too many photographers.

RATTLING. More than once during the peak of the rut, we have attracted whitetail bucks by rattling deer antlers together. That tactic is very well known. Not so well known is the fact that the rattling of heavier antlers will also attract heavier mule deer, elk, and occasionally moose. Lacking the antlers, it is even possible to make any of the antlered males show themselves by rattling a dead branch in dry brush or against a sapling. I've done this on occasion to cause an animal to move into sunlight just enough to allow a better composition. I also once located a fine bull elk, that had been totally hidden, when I shook a small lodgepole pine.

During the annual rut, I've often rattled up bucks and bulls by using antlers shed by bucks to simulate a fight.

CALLS. Various calls will work, depending on the time of year and skill of the user. With a little practice anyone can blow one of the commercial elk bugles well enough during the rut to make a bull respond. But once an animal knows you've fooled him, you'll waste your time

Murry Burnham, famed Texas game caller, calls adult whitetails by imitating the bleat of a fawn.

Using a rolled-birchbark megaphone, a guide makes grunting sounds that attract rutting moose.

trying to call him again. Canadian moose in rut respond to grunting through a megaphone made of rolled up birch bark or roofing shingle material. I know of no calls that work with any consistency on mule deer, but my old friend Murry Burnham of Marble Falls, Texas—the greatest game caller of all—can attract whitetails almost year-round with his manufactured mouth calls.

SCENTS.   We have had varying success with scents, specifically with whitetail deer scents. With varying labels, these are sold as "buck scents," "buck lure," "sex scent," and "doe-in-rut buck lure," among others. We have used scents often, carefully following instructions, but most often bucks have shown only passing, if any, interest. There may have been a couple of times when chemical buck lure might have helped keep a large male within our camera range.

During a whitetail filming expedition, we located an area where at least two large bucks, judging from hoofprints, had made their characteristic ground scrapes near our photo blind. This was an open hunting area and the season was in full swing. Mostly as an experiment, I made several false scrapes of my own to suggest that another buck claimed the territory; and I buried nearby two cotton swabs doused with doe-in-rut lure. That evening, a splendid 11-point buck came suspiciously to the spot. But on finding the scrape and smelling the lure, he stood for several minutes seeming completely puzzled. He didn't even react to the noise of my motor drive on that still evening. To this day, that buck remains among the best whitetail trophies we have photographed.

Dangers of Photography

Is big-game trophy hunting with a camera dangerous? Compared to driving a car, boating on a lake, or walking down some city streets at night, the answer is *No*. But big-game photography holds an element of danger that every would-be camera hunter should be aware of. All large animals have the potential to injure or even kill a person. Those that live in sanctuaries are more likely to cause trouble than those in areas where hunting is allowed because they may at times be less fearful of man than animals that are totally wild.

All of the horned and antlered animals in this book are products of what biologists call *photoperiodism*. Their lives are largely governed by how much daylight exists through the seasons. Each mammal's eye functions like the built-in exposure meter on a camera. It registers the amount of light, simultaneously transmitting the information to the hypothalamus of the animal's brain, and this either stimulates or retards activity in the pituitary gland. Photoperiodism governs when caribou begin their migrations, which is at roughly the same time every year. It also governs when the rut begins for the various species, also at about the same time every autumn. The rut of any species is the most exciting time to go trophy hunting. It is also the time when the males are least wary and most dangerous. But which species bear the most careful watching?

Pronghorn and caribou are the least dangerous of the animals discussed in this book. I have never heard of either to act the least bit aggressively toward people.

Even when infected with the high fever of the rut, mule deer seldom act mean toward people or other animals except rival bucks. But nonetheless, we always keep an eye on rutting bucks and stay a safe distance away. We are especially careful about frustrated bucks that have come out second best in fights with stronger bucks. Mule deer should never be taken for granted.

In our experience, moose have remarkable tolerance of people photographing them during the rut. Only once have I seen a moose launch a short charge toward a photographer who had advanced near enough to almost poke the camera in the bull's face. The bull could have gored and trampled the man, who fell trying to escape, but the bull did not. Moose deserve respect for their great size alone, but they are also faster afoot than seems possible for their ungainly appearance. We give moose plenty of room.

It is miraculous that photographers have not yet been killed by rutting bull elk in Yellowstone Park. Any large male defending his harem is a picture of mixed fury and frustration because he is constantly being harassed and challenged by rival bulls. Fortunately, bulls usually take out their frustrations by digging up clods of earth and shredding small trees in their path, instead of shredding the photographers who venture much too close for their own safety. The American bull elk in the rut is definitely among the species not to be trifled with, especially if you are away from climbable trees.

I once saw an elk drive its antlers through the bed of the pickup truck of the rancher who had raised it. This occurred while the rancher was trying to deliver hay to feed the animal. Peggy and I were sitting in the pickup cab at the time. No wonder we always keep a solid barrier between ourselves and large mammals that are not truly wild.

We have not had enough experience with muskoxen to be able to offer a reliable comparison of their aggression with that of other large mammals. But biologists we've met who have studied wild muskoxen as well as managers who have worked with captives, consider them unpredictable at best. Captives, even with plenty of room to roam, can be very aggressive.

Neither wild goats nor any of the wild sheep have ever made aggressive moves toward Peggy or me, even though we have spent lots of time photographing them in a variety of situations. But I have seen goats and sheep make threatening gestures toward people who violated what the animals felt was safe distance. These threatening gestures are the same as are made to rival animals: turning head and body sideways, then tossing the head backward. Although I feel no apprehension when among goats and sheep, I would never position myself on a cliff or sheer mountainside where a sudden nudge or lunge could send me over the edge. Nor would I dispute a large billy goat's right to pass along a narrow trail. I wouldn't want to appear to challenge his position on a feeding area, either.

BISON HAZARDS.   Knowing the threatening signals, and then backing off, can keep you out of trouble. And speaking of trouble, the seemingly ponderous, slow bison can cause plenty of it faster than a human eye can follow. Getting too near a bull bison is simply asking for a trip to the hospital or the funeral parlor. In 1982, in Yellowstone Park alone, eight people trying to photograph bison were gored or tossed by buffalo. By 1985 there were 25 injured, with several gravely hurt, and two people were killed. It's a wonder to me that the toll has not been greater. Peggy and I have seen photographers creep within a few yards of old solitary bulls and cows with calves—the two most dangerous categories—for little reason except bravado.

Most often a bison shows irritation by arching its back and licking its nose and face. (Licking the nose and face is also a sign elk and moose use to show displeasure.) Bull bison also may paw the ground and roll bloodshot eyes before coming after you. But just as often, you may be impaled by only a nonchalant toss of the head and horns, after which the bison continues to graze as if nothing happened. The best advice is to always give bison a slightly wider berth than seems necessary.

GRIZZLY HAZARDS. Photographers in Yellowstone, Glacier, Denali, and all other national parks of western Canada and Alaska are working in grizzly bear country. There you are likely to encounter bears anytime, anywhere, near the roads or on backcountry trails. Only a foolish photographer fails to keep watch for them. I have not found black bears to be aggressive, but I like to know of their presence anyway.

Grizzly bears do not really seek confrontations. But photographers absorbed with picture taking sometimes come upon grizzlies suddenly and unexpectedly. In Alaska, I once almost stumbled upon a bedded grizzly because I had become too engrossed in photographing Dall sheep. The grizzly ran away immediately, but the next one I meet like that might not. Peggy and I have also come far too close to grizzlies when shooting elk near the Dunraven Pass road in Yellowstone. Photographing and coping with bears is covered in much detail in *Erwin Bauer's Bear in Their World.*

A prime hazard with grizzlies occurs when you surprise them with your presence. Entering within a grizzly's programmed fight-or-flight distance leaves the outcome entirely up to the grizzly.

THE MOST DANGEROUS ANIMAL.   Which, then, is the most dangerous North American game animal of all? My vote goes to the whitetail buck that has been raised either in captivity or in some sanctuary where he has become accustomed to people. Whitetail bucks are not so much unpredictable as they are inclined to be aggressive during the rut. The Bambi of summer strikes out with both antlers and cloven hoofs, without any of the provocations that might make a moose or even a bull elk attack. More people have been killed by whitetail deer than any other large animals except bears.

   Males during the breeding season are not the only horned and antlered animals to beware of. Although some, if not most, females with calves will quickly desert their young rather than face an advancing human, many can just as quickly become more terrible and aggressive protectors than any bulls or bucks. Keep a safe distance from cows that lick their chops while flattening their ears and raising the hair on their necks and shoulders. When this happens you should tactfully retreat and switch to a longer telephoto lens.

An average whitetail buck measures about 15 inches across the chest and 15 inches from ear tip to ear tip. Using either of those references, you can figure that a buck whose antlers spread several inches beyond the 15 on each side and sport many long, symmetrical tines is usually in the trophy class.

An Addict's Wish

Peggy and I have photographed wild trophies all over North America. Many of the finest trophies in our files are of truly awesome animals. But the best thing about this game, this addiction, is that we can keep on hunting as long as film is available and the old legs hold out. I hope that is forever.

Where To Find Them
By Peggy Bauer

THE UNITED STATES

Northeast

DELAWARE is not widely known for its large mammals, but **Prime Hook National Wildlife Refuge** has whitetails (*RD 1, Box 195, Milton, DE 19968*).

MAINE is home to a population of whitetails in **Acadia National Park**. The fire of 1947 destroyed the evergreen forest. The new growth of quaking aspen, pin cherry, and birches makes good deer forage (*RFD 1, Box 1, Bar Harbor, ME 04609*). **Moosehorn National Wildlife Refuge** at the easternmost tip of the U.S. is home to a large number of mammals including some whitetails (*Box X, Calais, ME 04619*).

MARYLAND isn't a prime area for large wildlife, but there are some whitetails at **Blackwater National Wildlife Refuge**. A small herd of sika deer was planted there too (*Rt. 1, Box 121, Cambridge, MD 21613*). There are also whitetails at **Eastern Neck National Wildlife Refuge** (*Rt. 2, Box 225, Rock Hall, MD 21661*).

NEW JERSEY has a good population of whitetails in the rural areas, but the best public place to view them is at the **Great Swamp National Wildlife Refuge**, where in the past they have proliferated to such an extent that they endangered both their habitat and themselves (*RD 1, Box 148, Basking Ridge, NJ 07420*).

NEW YORK has whitetail deer at some privately owned reserves and at **Iroquois National Wildlife Refuge** (*RD 1, Casey Rd, Bason, NY 14013*) and at **Montezuma National Wildlife Refuge** (*RD 1, Box 1411, Seneca Falls, NY 13148*), which has had some albino individuals, too.

PENNSYLVANIA has some good private preserves scattered throughout the state (inquire locally) and a population of whitetails at **Erie National Wildlife Refuge** (*RD 2, Box 191, Guys Mills, PA 16327*).

VIRGINIA is one of the better places in the region to find wildlife, especially whitetail deer. In addition to private game areas, **Shenandoah National Park** is a fine place to look for deer (*Rt. 4, Box 156, Luray, VA 22835*). There are also three national wildlife refuges that all have a whitetail population: **Chincoteague**, which also has some sikas (*Box 62, Chincoteague, VA 23336*); **Mason Neck** (*9502 Richmond Highway, Lorton, VA 22079*); and **Presquile** (*Box 620, Hopewell, VA 23860*).

Southeast

SOUTH CAROLINA's Carolina Sandhills National Wildlife Refuge, although best known for it's red cockaded woodpeckers, is also a good place to look for whitetail deer beside the roads in late winter and early spring (*Route 2, Box 130, McBee, SC 29101*).

FLORIDA's single national park, **Everglades**, while a wonderful place, is not the best for whitetail deer. Still, they do exist and can be seen there (*Box 279, Homestead, FL 33030*). **Florida Keys National Wildlife Refuge** is the place to see the tiny key deer, a subspecies of whitetail. A bragging buck measures about 28 inches at the shoulder and a fawn less than a foot. Nearly extinct when the refuge was established in 1954, key deer are now plentiful and best seen on Big Pine and No Name keys from April to August (*Box 510, Big Pine Key, FL 33043*). **St. Vincent National Wildlife Refuge** stocked exotic species, then thought better of it and removed all but a few sambar deer (from India and Sri Lanka), which remain and have thrived. There are also some whitetails (*Apalachicola, FL 32320*).

GEORGIA has many whitetails, and a good number are in **Piedmont National Wildlife Refuge** (*Round Oak, GA 31080*). In addition, **Stephen Foster State Park**, located near Okefenokee Swamp at the southern border, is excellent for whitetails (*Fargo, GA 31631*).

NORTH CAROLINA shares a common border with Tennessee and astride this line lies **Great Smoky Mountains National Park**. Once wolf, elk, and bison lived there; now these animals are gone, but whitetails remain and can easily be seen at Cade's Cove, an 11-mile loop road, from September to January (*Gatlinburg, TN 37738*).

Mid-South

The mid-South is not our best wildlife area for larger mammals, though **ALABAMA's Eufaula National Wildlife Refuge** has some whitetails that can most easily be seen along the roadsides in early morning (*Rt. 2, Box 97B, Eufaula, AL 36207*).

ARKANSAS has quite a good number of whitetails, and the best public places to see them are in three of the national wildlife refuges: **Felsenthal** (*Box 279, Crossett, AR 71635*); **Holla Bend** (*Box 1043, Russellville, AR 72801*); and **White River**, where black bear might also be seen (*Box 308, DeWitt, AR 72042*).

LOUISIANA is a wonderful area to find birds and other smaller creatures, but poor for larger mammals. Whitetail deer, however, can be found in the following four national wildlife refuges: **Catahoula** (*P.O. Drawer LL, Jena, LA 71342*); **D'Arbonne** (*Box 3065, Monroe, LA 71201*); **Delta-Breton** (*Venice, LA 70091*); and **Lacassine** (*Rt. 1, Lake Arthur, LA 70549*).

MISSISSIPPI, like its neighbors, is poor in larger game, but there are whitetails in two national wildlife refuges: **Hillside** (*Box 107, Yazoo City, MS 39194*); and **Yazoo** (*Rt. 1, Box 286, Hollandale, MS 38748*).

TENNESSEE shares a border with North Carolina and also shares **Great Smoky Mountains National Park**. Whitetails can be seen in two national wildlife refuges: **Hatchie** (*Brownsville, TN 38012*); and **Tennessee** (*Box 849, Paris, TN 38242*).

Great Lakes States

Most of this area of the country is used so relentlessly for agriculture and dairy farms that little is left for wildlife. Even the fencerows that once offered shelter are now gone in many places.

ILLINOIS has populations of whitetails at two national wildlife refuges: **Crab Orchard** (*Box J, Carterville, IL 62918*) and **Mark Twain**, a large area consisting of a dozen units (*RR #2, Havana, IL 62644*).

INDIANA, also intensively cultivated, has a single national wildlife refuge that contains whitetails: **Muscatatuck** (*Box 631, Seymour, IN 46274*).

MICHIGAN's Isle Royale National Park is a scenic area where moose can be seen in a wonderful aquatic setting (*Houghton, MI 49931*). Albino whitetails have been seen with increasing frequency on the Upper Peninsula. Whitetail deer live in **Seney Na-**

tional Wildlife Refuge near Great Manistique Swamp on the Upper Peninsula (*Star Route, Seney, MI 49883*). They also find a home in **Shiwassee National Wildlife Refuge** (*6975 Mower Road, Saginaw, MI 48601*).

MINNESOTA's Voyageurs National Park is largely aquatic, but is home to a few moose and whitetail deer (*P.O. Box 50, International Falls, MN 56649*).

By far the most rewarding national wildlife refuge (and there are many in Minnesota) is **Agassiz**. Moose calves are dropped in early May, and three weeks later the deer drop their fawns, coinciding with the first flights of waterfowl. There are also some elk, but these are better seen on adjoining and nearby state and private lands. The best time to see the moose and deer is in the spring when calves and fawns are born, and in fall when antlers reach their greatest growth and the velvet is shed (*Middle River, MN 56737*).

There are four other national wildlife refuges where whitetails can be seen: **Big Stone** (*25 Northwest Second St., Ortonville, MN 56278*); **Rice Lake** (*RR No. 2, McGregor, MN 55760*); **Sherburne** (*Rt. 2, Zimmerman, MN 55398*); and **Tamarac** (*Box 66, Rochert, MN 56578*).

There is also an area of lands owned by the states of Iowa, Minnesota, Wisconsin, and Illinois called **Upper Mississippi River Wildlife and Fish Refuge**, which can be seen from Great River Road or, better yet, by boat. Some of the refuge's whitetails can be seen swimming from island to island (*Minnesota headquarters address: 122 West Second St., Winona, MN 55987*).

WISCONSIN, with no national parks, has several national wildlife refuges, two of which contain whitetail deer: **Horicon**, best known for its huge population of Canada geese (*Rt. 2, Mayville, WI 53050*), and **Trempealau** (*Rt. 1, Trempealau, WI 54661*).

North-Central States

IOWA, better known for hogs and corn, has two national wildlife refuges that contain whitetails: **DeSoto**, where as usual they are most easily seen early and late in the day (*Box 114, Missouri Valley, IA 51555*), and **Union Slough**, where whitetails can be viewed from the trails and roads and at the south end near Rt. B-14. In winter they congregate in groups ("yard up") in interior areas nearby (*Rt. 1, Box 32-B, Titonka, IA 50430*).

KANSAS, so intensively cultivated as to leave little space for wildlife outside designated areas, has two national wildlife refuges that contain whitetails: **Kirwin**, which has both whitetails and mule deer as well as badgers and a prairie dog town (*Kirwin, KS 67644*), and **Quivira**, where there are also prairie dogs and some burrowing owls (*Box G, Stafford, KS 67578*).

MISSOURI has two national wildlife refuges with populations of whitetail deer: **Mingo**, where the deer can be seen from auto, boat, or on foot (*Box V, Puxico, MO 63960*), and **Swan Lake**, which has a large herd of deer but no swans (*Box 68, Sumner, MO 64681*).

NEBRASKA, true to its agricultural heritage, has left little room for its native wildlife. But it does have two national wildlife refuges, Ft. Niobrara and Crescent Lake. **Ft. Niobrara** is best. Elk, bison, pronghorn, and both whitetail and mule deer live here, as well as some longhorn cattle. The bison drop calves in May and the deer produce young in June. As always, the mule deer are easier to approach than the whitetails. Bison create a great stir when they rut in July, while the pronghorn, deer, and elk ruts begin in September. In the fall, too, there is an exciting bison roundup (*Hidden Timber Star Rt., Valentine, NE 69201*). **Crescent Lake** is home to pronghorn and both mule deer and whitetails (*Star Rt., Ellsworth, NE 69340*).

NORTH DAKOTA's Theodore Roosevelt National Park is the best place in the state to see and photograph horned and antlered animals. Bison are impossible to miss, pronghorn wander in good supply, and both whitetail and mule deer inhabit this rather large preserve. The sights are good any time of the year in the Medora

section, with autumn the preferred season (*Box 7, Medora, ND 58645*).

The second-best area in North Dakota for wildlife viewing is **Sully's Hill National Game Preserve**. Don't come in winter when weather makes the roads impassable. But you'll find bison at any other time, including cows with their calves when the young are only a few days old. Whitetail deer and elk are out of view during the last week in May and early in June when they retreat to drop their young, but otherwise they are easy to see. The bison rut takes place in early fall, followed by that of elk a bit later in the season (*Ft. Totten, ND 58335.*)

The following national wildlife refuges also have some whitetail deer: **Arrowwood** (*Pingree, ND 58476*); **Audubon** (*Coleharbor, ND 58531*); **DesLacs** (*Box 578, Kenmare, ND 58746*); **J. Clark Salyer** (*Box 66, Upham, ND 58789*); **Tewaukon** (*RR 1, Cayuga, ND 58013*); and **Upper Souris** (*RR 1, Foxholm, ND 58738*).

SOUTH DAKOTA has two valuable national parks, Wind Cave and Badlands. Created as a national monument in 1939, **Badlands** became a park in 1978, and when the plowing stopped the native grasses returned and with them, the antelope. In 1963 more were reintroduced and have prospered. Look for them in the park's Sage Creek Wilderness Area. The area's Audubon subspecies of bighorn sheep is long extinct, but in 1964 another population of bighorns was introduced and survives today (*Box 6, Interior, SD 57750*).

Wind Cave National Park, in addition to the caves, has wonderful horned and antlered wildlife. The pronghorn are especially confiding, mule deer are easily seen as are the elk, and the bison, although less visible, are there nonetheless (*Hot Springs, SD 57747*).

Custer State Park is one of the best state parks in the U.S. Here, where prairie and mountains meet, you'll find deer, bison, pronghorn, elk, and perhaps even bighorns (*Hermosa, SD 57744*).

Whitetail deer exist on four of South Dakota's national wildlife refuges: **Lacreek**, where both whitetail and mule deer can be seen from the road (*South Route, Martin, SD 57551*); **Lake Andes**, where both species can occasionally be seen from the road (*Box 279, Lake Andes, SD 57336*); **Sand Lake**, where fawns are numerous in early summer and, incidentally, where badgers are less cautious than usual (*RR 1, Columbia, SD 57443*); and **Waubay** (*RR 1, Box 79, Waubay, SD 57273*).

Rocky Mountain States

COLORADO, a state of great natural beauty, has managed to maintain much of its original character and most of its wildlife population. Two good places to see some of it are in Mesa Verde and Rocky Mountain National Parks.

Mesa Verde is best known for its cliff-dwelling ruins, but less well known is the fact that it also supports mule deer in good numbers. The deer often stand in wonderful, scenic spots (*Mesa Verde National Park, CO 81330*).

Rocky Mountain, at a high elevation astride the Continental Divide, has a nice variety of horned and antlered wildlife, though they are not particularly easy to see. Look for mule deer, elk, and bighorn sheep, preferably from horseback (*Estes Park, CO 80517*).

Alamosa National Wildlife Refuge has herds of both mule deer and elk. They can be seen from the bluff beside the refuge road on the meadow below (*Alamosa, CO 81101*).

Arapaho is another wildlife refuge worth visiting. At 8,300 feet, it's the highest outside Alaska, but the low percentage of oxygen doesn't seem to bother the pronghorn or the mule deer, which are easiest to see in winter. Look for elk, too (*Box 457, Walden, CO 80480*).

Browns Park National Wildlife Refuge, with a climate rather mild for Colorado, has populations of pronghorns and mule deer. The deer occasionally swim the river (*Greystone Rt., Maybell, CO 81640*).

IDAHO, exerting pressure against the great forces for timbering and mining, cattle- and sheep-ranching, has managed to save some of its natural beauty and wildlife. Three national wildlife refuges lie in the Gem State, and although officials have been pressed to allow domestic animal grazing within their borders, the refuges still manage to hold populations of wilderness monarchs.

Camas, named for the lovely lily of the area, has some mule deer that are easiest to see early and late in the day. Occasionally pronghorn are seen and even more occasionally, moose *(Hamer, ID 83425)*. **Gray's Lake**, best known as the nesting place for a new population of whooping cranes, also has some moose and mule deer. It's not the best place to see them, however *(Box 837, Soda Springs, ID 83276)*. Finally, **Kootenai**, just south of the Canadian border, has both mule and whitetail deer. Infrequently moose and elk are seen, too *(Star Rt. 1, Box 160, Bonners Ferry, ID 83805)*.

Several state parks have small numbers of larger mammals. **Farragut** has whitetails and occasional elk *(Box F, Athol, ID 83801)*. **Priest Lake**, high on the panhandle, has both whitetails and mule deer, in addition to some moose and elk. Very rarely, mountain goats are seen *(1838 Lincoln Way, Coeur d'Alene, ID 83814)*. **Three Island Crossing** lies on the Oregon Trail where ruts made by wagon wheels can still be seen. There, you'll find bison herd in a fenced pasture. Longhorns, although technically not "wildlife," are also there *(Box 609, Glenns Ferry, ID 83623)*.

Although there are no national parks within the borders of Idaho, it is possible to see other parts of the state using private guides or outfitters. Hells Canyon, the gorge of the Snake River between Idaho and Oregon, is constantly under siege by those with plans to decimate it, but thus far can be seen in its natural, marvelous condition. Whitewater float trips are conducted by many organizations. Write *Idaho Outfitters and Guides Association, Box 95, Boise, ID 83701* for a catalog of upcoming trips. Look for bighorn sheep from the rafts.

For similar trips on the Salmon River and the Middle Fork, which is bighorn sheep and mule deer country, write: *American River Touring Association (ARTA), 445 High Street, Oakland, CA 94601*.

Shepp Ranch, on the banks of the Salmon River, offers accommodations for those who choose to hike or ride horseback to find mule deer, elk, and other wildlife of the area *(Shepp Ranch, Box 3417, Boise, ID 83703)*.

MONTANA, with Wyoming, is the premier state for the horned and antlered mammal enthusiast. **Glacier National Park**, hard on the Canadian border, is a magnet for wildlife lovers who flock here in great numbers from earliest spring through fall. With the sole park road closed in winter, there is little access at that time. Mule deer, bighorn sheep, elk, moose, and mountain goats exist there in healthy numbers *(West Glacier, MT 59936)*. To see the goats at close range, hike from Lake McDonald to Sperry Chalet. They look best in fall, though their kids are appealing at any time. Bighorn sheep are usually near **Many Glacier Hotel**, a wonderful base camp for the naturalist. Reservations are available at the park or at Greyhound Tower *(Phoenix, AZ 85077)*. The sheep wander the hotel grounds in late fall.

Three of Montana's many national wildlife refuges have whitetail deer. **Bowdin** also hosts pronghorn, and young are often grazing beside the tour route *(Box J, Malta, MT 59538)*. **Medicine Lake** has the largest number of whitetails in the northeastern part of the state. Pronghorns are infrequently seen *(Medicine Lake, MT 59247)*. **Metcalf** (once called Ravalli) has a large herd of whitetails and other, smaller creatures *(Box 257, Stevensville, MT 59870)*. **Red Rocks** on the Idaho border is best known for its trumpeter swans but also has moose, mule deer, and pronghorn. The refuge road is usually closed in winter, and these creatures might better be seen in other, nearby areas *(Monida Star Rt., Box 15, Lima, MT 59739)*. **C.M. Russell** has populations of pronghorn, mule and whitetail deer, and some bighorn sheep *(Box 110, Lewistown, MT 59457)*.

The National Bison Range can be an exciting destination. Herds of huge bison roam at will, dropping calves in early spring, mating in August, and suffering a roundup in early September. Elk are evident all year and rut in late September, pronghorn are easiest of all to photograph, and both whitetail and mule deer can be

found at the end of each day. Bighorns are least visible, but not impossible to find *(Moiese, MT 59824)*.

There are also many private concerns able to guide the visitor toward wildlife others may miss. Usually these services provide horseback trips geared for the dude with everything supplied. One such concern is **Miller's Wilderness Outfitters** *(Box 1083, Cooke City, MT 59020)*.

UTAH may be the most scenic state in the union. Fortunately it also has the most national parks. Most of these have achieved fame because of their marvelous geographical features, not their masses of large wildlife. However, while admiring nature's handiwork, be alert for mule deer in four of the national parks: **Bryce Canyon** *(Bryce Canyon, UT 84717)*; **Canyonlands** *(446 South Main St., Moab, UT 84532)*; **Capital Reef** *(Torrey, UT 84775)*; and, biggest and best, **Zion**, where the deer are easiest to see in Zion Canyon in November and December *(Springdale, UT 84767)*.

Two of the state's national wildlife refuges also have some modest numbers of horned and antlered game: **Fish Springs**, where some mule deer and pronghorn live out their lives *(Dugway, UT 84022)*, and **Ouray**, which also has some muleys and, on the east side, pronghorns *(447 East Main St., Vernal, UT 84078)*.

WYOMING is home to the most exciting large creatures in the Lower 48. The destination of wildlife lovers, the Cowboy State has two of the most outstanding national parks in the country. **Grand Teton** contains good numbers of elk, bison, moose, mule deer, and pronghorn. Look for elk in most open meadows, especially early and late in the day. The flats surrounding Timbered Island are best. Mule deer does and their fawns graze beside alpine trails, but the bucks remain high and hidden until the fall rutting season. Moose wander high (as well as beside the lakes), but are easy to find. Pronghorns are always on flat, open sage meadows, often the same place as the elk, near Timbered Island. The small band of bison continues to grow and wanders unpredictably, often where they are less than welcome. Most roads are closed in winter *(Moose, WY 83012)*.

Yellowstone National Park, America's first and largest, outside Alaska, is as well known for its wildlife as for the renowned thermal features such as Old Faithful. Summer is the poorest season for wildlife viewing, but even then the early riser will not go unrewarded. Elk are common around Mammoth Campground and their ever-increasing numbers may account for the fact that mule deer are seen less frequently here than in years past. On frosty fall mornings, often even before the first glimmer of light, bull elk bugle challenges to all comers. Sparring between equally matched bulls is frequent. Mule deer are more visible at the north end of the park later in the fall and during winter near all the thermally heated areas where grasses continue to grow in the warmed earth. Firehole Flats and the Hayden Valley are good places to find bison, although they are becoming so numerous that they can now be found where they never ventured before. Moose are frequent visitors at Hayden Valley in summer and at Pelican Creek in the fall. Pronghorns are almost always near the north entrance, while bighorn sheep conduct ancient mating rituals on the grassy plateaus over the Gardiner River in late November. Sometimes the crash of bighorn rams in combat echoes the ring of elk fighting on an adjacent ridge.

A winter visit to Yellowstone is an unforgettable experience. The wildlife is tolerant of humans then, and the snow-covered landscapes dotted by steaming thermal pools of turquoise are a glittering wonderland. For information on excursions contact: *Yellowstone Park Division, TW Services, Inc., Yellowstone National Park, WY 82190*.

The National Elk Refuge *(Jackson Hole, WY 83001)* is a spectacle of over 5,000 animals each year from late November and usually until April, when the last of the creatures wanders off to forage on the new growth. Elk bulls maintain their huge antlers until early spring, so they can easily be photographed with the Teton Range of the Rockies as a magnificent background. Horse-drawn sleighs take visitors to within close range of the elk (wapiti). Mule deer are less easy to see, but many live on the refuge's buttes during winter. Recently the enlarged herd of bison from **Grand Teton National Park** *(Moose, WY 83012)* has spilled over onto the refuge and freely partakes of the feed distributed for the elk. The bison have a tendency to gore elk that venture too close to the feed and present a growing problem to refuge personnel.

Pronghorns and mule deer are fairly easily seen at **Seedskadee National Wildlife Refuge** in the southern part of the state *(Box 67, Green River, WY 82935)*.

Hot Springs State Park is a good place to photograph free-roaming bison and their relatively small rust-colored calves in June *(Thermopolis, WY 82443)*.

Whiskey Basin, east of Dubois in western Wyoming, is famous for its population of bighorn sheep that are easy to approach in winter when the strong winds blow the snow from high slopes. It is just off U.S. 287 *(Dubois, WY 82513)*.

There are also private businesses, usually small and family-run, which will take visitors to beautiful areas where wildlife is known to roam. Usually such trips are by horseback on dude-wise mounts with everything furnished. One concern is **Crystal Creek Camps** *(Star Route, Box 44-A, Jackson, WY 83001)*. Another is **Ken Clark** *(Rt. 1, Afton, WY 83110)*.

Southwest

ARIZONA, an extremely dry state, is also a fragile area whose wildlife has been hard put to withstand the onslaught of human numbers and destructive intruders such as wild horses and burros. These continue to destroy the range and foul the water, yet somehow receive the well-meant but misplaced sympathies of uninformed wildlife devotees.

Grand Canyon National Park, where the Colorado River has etched a monumental gash on the face of the earth, has little wildlife, but mule deer can be seen on the North Rim especially from late fall until freeze-up *(Box 129, Grand Canyon, AZ 86023)*.

Saguaro National Monument has some desert mule deer that can be seen in the Rincon Unit, which is higher and has somewhat more rainfall than the remainder of the monument. There are also some whitetail deer *(Box 17210, Tucson, AZ 85731)*.

Cabeza Prieta National Wildlife Refuge also manages **Kofa**, a sister refuge. There are some desert bighorns here, and the extremely endangered Sonoran pronghorns can be seen near a water hole at the east end of the area. Roads require a 4WD vehicle and a permit is required for entry *(Box 1032, Yuma, AZ 85364)*.

NEW MEXICO looks in places as if the overgrazing sheep have completely devastated the landscape. But there are several areas in the state of surprisingly lush beauty and respectable numbers of wild creatures.

Carlsbad Caverns is notable mainly for the wonderful sculptures and caves beneath the earth's surface. But be alert for the desert mule deer that roam the well-vegetated surrounding area. Smaller than their northern cousins, these deer are at home with prickly pear cactus and the other hardy plants that survive in the harsh area *(3225 National Parks Highway, Carlsbad, NM 88220)*.

Bandelier National Monument was once a fine place to find desert mule deer, but widespread destruction by feral burros has changed things for the worse *(Los Alamos, NM 87544)*.

Bosque del Apache National Wildlife Refuge in the center of the state is a prized jewel in the refuge crown. Most visitors come to see the young whooping cranes as well as the huge numbers of other cranes, geese, and ducks, but there is also a large population of mule deer that are reasonably easy to see and photograph from the refuge tour road. Some very large bucks have been admired here *(Box 1246, Socorro, NM 87801)*.

For absolutely certain viewing of the wildlife of this desert area, visit the **Living Desert Zoological and Botanical Park** 30 miles north of Carlsbad on the main highway. There is a large hoofed animal enclosure. In a separate area we found a cow buffalo and her calf, about ten minutes old *(U.S. 285, Carlsbad, NM 88220)*.

NEVADA is far from our most productive state for finding monarchs of the wilderness, but there are three national wildlife refuges where some creatures can be found. **Desert**, where all roads are primitive and a 4WD vehicle is required, has some desert bighorn sheep and desert mule deer *(1500 No. Decatur Blvd., Las Vegas, NV 89108)*. **Ruby Lake**, a scenic area with many watery impoundments, has a population of mule deer that are occasionally preyed upon by cougars *(Box 649, Elko, NV 89801)*. Best though still not highly recommended, is **Sheldon Antelope Refuge**. On the Big Springs Table where the weather is less severe than elsewhere on the refuge, up to 8,000 pronghorns might be found in winter. Young are produced in May and herds of 100 gather in summer together with mule deer. There are a few bighorns and, unfortunately, burros. This is a wild and remote area where viewing is difficult *(Denio, NV 89404)*.

Hoover Dam, at the southeastern corner of the state, backs up waters of the Colorado River to form Lake Mead. Bighorn sheep can usually be seen on the hillsides near the lake *(Lake Mead National Recreation Area, 601 Nevada Hwy., Boulder City, NV 89005)*.

OKLAHOMA has three national wildlife refuges, each with a population of whitetail deer: **Salt Plains** *(Rt. 1, Box 76, Jet, OK 73749)*; **Sequoyah**, *(Box 398, Sallisaw, OK 74995)*; and **Tishomingo** *(Box 245, Tishomingo, OK 73460)*. Best of the public areas for wildlife viewing is **Wichita Mountains National Wildlife Refuge**. The first area set aside for a mammal, the bison, Wichita now has a good number of these hulks that are easy to see in attractive surroundings. Elk have been reintroduced, and the refuge holds tours beginning each September to see the great antlered animals in the rut when they bugle, gather harems of cow elk, and challenge other bulls that might consider invading the female herd. Although not considered native, longhorn cattle also roam here *(Box 448, Cache, OK 73527)*.

TEXAS's Big Bend National Park, nestled in the elbow of the Rio Grande River like a fat dog in a hammock, is a lightly visited area better known for its scenic features than its monarchs of the wilderness. But there are desert mule deer here that are reasonably confiding, as well as some pronghorns. For many of us the most interesting creature is the tiny subspecies of whitetail deer, the Carmen Mountain whitetail *(Big Bend, TX 79834)*.

Guadalupe Mountains National Park is newly designated as a park and almost totally undeveloped. Nonetheless, it's worth a visit in winter for the scenery and the mule deer that are quite easy to find *(Write: % 3225 National Parks Highway, Carlsbad, NM 88220)*.

Aransas National Wildlife Refuge is the best public place in Texas to photograph whitetail deer. There are some bragging-size bucks here that many have seen and photographed from the tour road. Aransas is best known as the place where one can see the greatly endangered whooping crane from a commercial boat *(Box 100, Austwell, TX 77950)*.

Buffalo Lake National Wildlife Refuge also has some mule deer on the sanctuary, but with water sources scarce, visits are probably best preceded by an inquiry *(Box 228, Umbarger, TX 79091)*.

Two other national wildlife refuges have some whitetail deer: **Hagerman**, where a photographer must constantly avoid the oil wells *(Rt. 3, Box 123, Sherman, TX)*, and **Laguna Atascosa** *(Box 2683, Harlingen, TX 78550)*. These last two areas are not enthusiastically recommended.

The Y.O. Ranch raises native mammals for hunting and also has numbers of non-native horned and antlered creatures. It offers photo safaris from an open jeep over the hundred square miles of the property. Hunted animals are never easy to photograph, however *(Box 3000, Mountain Home, TX 78058)*.

The Welder Wildlife Foundation has a beautiful large area set aside where groups are welcome to see the native wildlife. Whitetail deer are a specialty. A telephone call or letter is required first *(Drawer 1400, Sinton, TX 78387)*.

For a more complete listing of places to see and photograph wildlife in Texas, see our new book, *Photographing Wild Texas.*

Pacific Northwest

CALIFORNIA, as has often been pointed out, has everything. The wilderness monarchs, while not the state's main attraction, are included nonetheless.

Redwood National Park has some blacktail deer in addition to its huge, ancient trees. Blacktails, found only on the Pacific Northwest coast, are seen most easily here and in Olympic National Park in Washington. A subspecies of elk, the Roosevelt elk, is also seen in Redwood *(Drawer N, Crescent City, CA 95531)*.

Yosemite National Park, famous for its magnificent scenery, has some mule deer that gorge each season on the acorn crop in the park *(Yosemite National Park, CA 95389)*.

Point Reyes National Seashore is a grand seaside area where blacktail and fallow deer can easily be seen and photographed early in the day *(Point Reyes, CA 94956)*.

There are four national wildlife refuges in California with both horned and antlered creatures.

Klamath Basin, in both California and Oregon, occasionally has pronghorns, and mule deer are common. Take the tour road on Tule Lake and Lower Klamath units. Pronghorns are in the Clear Lake unit *(Rt. 1, Box 74, Tule Lake, CA 96134)*.

Modoc has just an occasional glimpse of pronghorns. A large herd may cross a corner of the refuge for two or three days only. Others can be seen in November and from March to May. Mule deer are a common sight in summer *(Box 1610, Alturas, CA 96101)*.

Sacramento has a population of blacktail deer *(Rt. 1, Box 311, Willows, CA 95988)*.

San Luis is one of the few places where one can see the rare Tule elk, the smallest of all North American elk. They are in a five-mile fenced enclosure *(Box 2176, Los Banos, CA 93635)*.

Tule elk can also be seen in the California State Preserve near Taft on I-5 north of Santa Barbara and west of Bakersfield.

The state also has a wonderful park with many blacktail deer on Angel Island in San Francisco Bay, which can be reached by ferry from San Francisco.

OREGON seems divided, north to south, into two vastly different areas: the green, wet western section along the Pacific Ocean, and the dry deserts of the eastern section. Crater Lake National Park in the Cascade Mountains that split the state is notable for its wondrous lake in the caldera of an extinct volcano. But the surrounding forests and meadows are home to mule deer that survive in magnificent scenic settings *(Box 7, Crater Lake, OR 97604)*.

Wilderness monarchs also exist in four national wildlife refuges in the state.

Hart Mountain, adjoining Sheldon Antelope Refuge in Nevada, is where the pronghorns spend the summer and drop their young. They can be seen every season but winter from the 20-mile tour route. There are also mule deer here and a few bighorn sheep *(Box 111, Lakeview, OR 97630)*.

Malheur, until the flooding of recent years, was the best place in Oregon for wildlife viewing. It is far less rewarding now, and the situation is not improved by the domestic cattle that are allowed to graze on refuge lands specifically set aside for wildlife. But you can still see mule deer from September to November in the meadows and at the edges of the willows *(Box 113, Burns, OR 97720)*.

The Willamette Valley Complex is home to a number of blacktail deer *(Rt. 2, Box 208, Corvallis, OR 97330)*.

Umatilla, in both Oregon and Washington, is a possible though not greatly recommended place to see mule deer *(Box 239, Umatilla, OR 97882)*.

WASHINGTON, like Oregon, is sliced in two by the Cascades, its western part wet and its eastern part dry.

North Cascades National Park is largely rugged and wet, although the eastern side is noticeably drier and warmer at all times. Columbian blacktail deer live here and can be seen from the trails and occasionally from the highway that transverses the park. Look too for mountain goats; they can often be found beside Lake Chelan in winter from a boat *(Sedro Woolley, WA 98204)*.

Olympic National Park is a very wet remnant of the temperate rain forest that once stretched for hundreds of miles along the Pacific Coast. Summer is the best time to visit, but the most difficult to find the Roosevelt, or Olympic, elk that thrive in the park. They descend in winter for better forage, and retreat to high meadows as the season warms. These elk are heavier and darker than their Rocky Mountain cousins. Blacktail deer can be seen on Hurricane Ridge, where they graze peacefully among masses of wildflowers *(600 E. Park Avenue, Port Angeles, WA 98362)*.

Mount Rainier National Park's main attraction is, of course, the mountain peak. But anyone interested in horned or antlered wildlife will find much in addition to the old volcano. The best place to find mountain goats is near Paradise, although there are also goats on Emerald Ridge, the Colonnades, and at Cowlitz Chimneys. Elk flourish on the Cowlitz Divide (reached by a steep trail) and also on Shriner Peak. They are best seen in autumn and move out of the park in winter. Most deer are blacktails, although some mule deer are found in the eastern portions of the park. They can be seen in the vicinity of Nisqually entrance station and then in dense forests as autumn approaches *(Mt. Rainier National Park, Longmire, WA 98397)*.

The Columbian White Tailed Deer National Wildlife Refuge was established to protect the small Columbian subspecies of whitetail. Fawns are easily seen in summer, and all of the deer are more visible in fall and winter when the vegetation dries. Look at the wooded edges from the tour road that circles the mainland section of the refuge. Elk arrive in mid-September to spend the winter *(Rt. 1, Box 376-C, Cathlamet WA 98612)*.

Ridgefield National Wildlife Refuge has fairly abundant blacktail deer *(Box 457, Ridgefield, WA 98642)*. And Turnbull National Wildlife Refuge claims that whitetails are common and easily seen *(Rt. 3, Box 385, Cheney, WA 99004)*.

Willapa National Wildlife Refuge is composed of three units; the Long Island unit of 6,000 acres is best for those interested in larger mammals. Access is by boat only, and of course one then travels by foot to see Roosevelt elk and blacktail deer. Incidentally, the densest population of black bears in this part of the country also lives here *(Ilwaco WA 98624)*.

The state of Washington manages many wild acres, three of which are of special interest: the Oak Creek and L.T. Murray Area, where hundreds of elk come in winter to feed at the headquarters on U.S. 12, White Pass Highway *(Regional Office, 2802 Naches Highway, Yakima, WA 98202)*; Sinlahekin Recreation Area, where in winter deer and bighorn sheep come to the lower elevations conveniently beside the road *(Regional Office, Box 1237, Ephrata, WA 98823)*; and the Wooter Wildlife Area, where bighorn sheep are seen year-round and elk in winter *(Regional Office, North, 8702 Division St., Spokane, WA 99218)*.

The Metropolitan Park District of Tacoma oversees wonderful Northwest Trek Wildlife Park, 35 miles from the city in Eatonville. A visit here is a must for anyone remotely interested in wildlife. Whitetail deer wander freely in the park, where you can see fragile-looking spotted fawns in spring. Elk, moose, and mountain goats are also on view in natural settings. Creatures other than the horned and antlered monarchs thrive here too, and all can be seen at their best *(Eatonville, WA 98328)*.

On the Olympic Peninsula is the Olympic Game Farm, a huge, drive-through public area where massive elk and exotic deer have the run of a large enclosed pasture. They are impossible to miss. Other creatures are here too, from the rare to the common, all easily seen and photographed. Follow the signs from Sequim *(Sequim, WA 98328)*.

Alaska

Nowhere in North America are the opportunities to see and photograph wildlife greater than in Alaska. The areas listed below do not begin to include all the possible places one might find horned and antlered creatures as well as other animals, but they are the most rewarding and easiest to reach.

Many consider Denali National Park (formerly Mt. McKinley) the best in America, and it may be just that. It is the easiest place in the U.S. park system to see barren-ground caribou and Dall sheep. Huge Alaskan moose are also easy to film. There are no deer in Denali, however.

The caribou are scattered during the summer months when most visitors come, and the animals can be seen in many areas singly and in twos or fours. By the end of August or early September, the velvet is rubbed from bulls' antlers and the newly hardened racks gleam rose red for several days. As the year progresses the caribou begin to gather in even larger groups as they prepare for the fall migration. The open woodlands east of Wonder Lake is a good place to seek them.

Dall sheep, the only pure white sheep in North America, stay high on grassy slopes in summer and are easy to locate from the road. They will not flee a careful, leisurely approach. Males and females remain separate in summer and meet for the fall rut.

Alaskan moose, visibly larger than the Shiras moose of the Rockies, range separately in summer in willow and alder, then begin to congregate early in September in Igloo Flats. They are wonderful subjects for camera lenses, binoculars, or the naked eye *(Box 9, Denali National Park, AK 99755)*.

Glacier Bay National Park, best known for its glacial activity, is home to a population of Sitka blacktail deer and some moose, too. This is not the best place to seek them, however *(Box 1089, Juneau, AK 99801)*. Another national park, **Katmai**, visited mainly because of its good salmon fishing and the presence of brown bears, also has some moose and caribou. They don't present a great spectacle, however *(Box 7, King Salmon, AK 99613)*.

Kenai Fjords was designated a national park to protect the marvelous congregation of sea life, both mammals and birds. A day's cruise on a boat run by Kenai Fjords Tour, Inc. *(Box 881, Seward, AK 99664)* slides the visitor to within close range of such sea mammals and birds as otters, seals, and puffins. Above this mass walk snow white mountain goats on steep dark cliff faces. Another section of the park is at Exit Glacier where massive Alaskan moose are often seen *(Box 1727, Seward, AK 99664)*.

The average person is not likely to get to **Kobuk Valley National Park**, but there are herds of caribou there and also some moose *(Box 287, Kotzebue, AK 99752)*.

Lake Clark National Park is a magnificent area even by Alaskan standards and can be reached by air only. Each elevation has its own wilderness monarchs: moose live in the green summer valleys, caribou roam the tundra, and Dall sheep graze the highest meadows *(701 C St., Box 6, Anchorage, AK 99513)*.

Wrangell-St. Elias National Park, America's largest, has Dall sheep that are easy to see. Mountain goats are here, too, but harder to find. At the north end in the Copper River valley and vicinity are Alaskan moose in the forest, tundra, and boggy lowlands *(Box 29, Glennallen, AK 99588)*.

The Portage Glacier area in the **Chugach National Forest** is popular with summer visitors, yet there is little or no wildlife seen at that time. Later in the season, however, mountain goats and Alaskan moose are easy to find *(Chugach National Forest Office, Anchorage Ranger District, P.O. Box 10-469, Anchorage, AK 99551)*.

Most people travel to **Kodiak National Wildlife Refuge** to watch for the great bears, but there are also many horned and antlered creatures. Three have been introduced: the blacktail deer, Roosevelt elk, and Dall sheep *(Box 825, Kodiak, AK 99615)*.

Yukon Delta National Wildlife Refuge encompasses a vast area of western Alaska including Nunivak Island, where muskoxen have proliferated so much that members of the herd have been transported to other areas. Nunivak is difficult to reach, but you can see muskoxen and reindeer year-round there *(Box 346, Bethel, AK 99559)*.

Chugach State Park, a vast mountainous area near Anchorage, has moose that can be seen at any time, and Dall sheep in view only from May to September *(Alaska State Parks, Chugach District, 2601 Commercial Drive, Anchorage, AK 99501)*.

There are also many private businesses that can show you the monarchs of their portion of the state. One is **Afognak Wilderness Lodge** on tiny Afognak Island, a satellite of Kodiak Island. In addition to the bears that feed on the spawning salmon, Afognak has an introduced population of Roosevelt elk and a large number of Sitka blacktails *(Seal Bay, AK 99697)*. Another is **Alaskan Wildlife Photography Adventures** *(Box 1557, Homer, AK 99603)*.

CANADA

There are no large game species that do not exist somewhere in Canada, with the great preponderance in the western part of the country. Once away from urban areas, the sparsely populated land has so much wilderness with wildlife that only a small portion can be listed below.

ALBERTA, together with British Columbia, has some of the most awesome mountain scenery on earth and also a large number of the mountain goats, bighorns, elk, and mule deer we associate with this topography.

Banff National Park is a perfect place to look for elk during the fall rutting season. In early mornings especially, look for open meadows near the roadsides where bulls collect their harems. You may hear them bugle from a long distance. Watch along the Bow River where the bulls often herd their cows across to the wooded islands and then to the far side as the sun rises. They may cross at the same place day after day. If you miss the best action one day, try again the next.

Deer are also more approachable in the autumn and can be seen from the network of trails or from roadsides. Look at mineral licks for wildlife (watch for the road signs); they are likely places for mountain goats. Just west of the Banff traffic circle on the main highway is a buffalo paddock where the creatures wander over 100 fenced acres. Moose are usually in the watery, low areas of the park. Even the townsite of Banff can be rewarding *(Box 900, Banff, Alta., CANADA T0L 0C0)*.

Jasper National Park is north of Banff, reached by the scenic Icefields Parkway. Banff and Jasper are similar. Here elk occasionally move in bands across the Athabasca River and are easy to see from many points along the road or from the 620 miles of trails in the park. Scenes are awesome with several peaks reaching over 10,000 feet. Mule deer are unwary in the park and can also be found on the outskirts of the town of Jasper, and occasionally in the wooded area near the railroad tracks. Bighorn sheep may be easiest of all to see close up. Ewes are often close beside the main roads and may cross the highway in front of large commercial trucks with complete confidence. Rams can often be seen high beside the roads and, if approached slowly and quietly, have little fear of human presence. Moose are found in damp areas. The park is crowded in summer, but much less so during the fall, which is the best time for the wildlife enthusiast to visit *(Box 10, Jasper, Alta., CANADA T0E 1E0)*.

Waterton Lakes National Park adjoins Glacier Park, in Montana. The parks are geologically similar, but in philosophy quite different. The Canadian park is more of a summer resort that features a golf course and motels with swimming pools. By taking the road to Red Rock Canyon, however, you'll often see bighorns, though their numbers have been decimated by disease in recent years. Elk are in the area, too, but they are hunted just outside the park boundaries and are extremely difficult to approach. In summer, drive through the buffalo paddock where there are plains bison. There are also mule deer beside the trails *(Waterton Lakes, Alta., CANADA T0K 2M0)*.

Wood Buffalo National Park is Canada's largest, extending into the Northwest Territories. Intended to protect the rare wood bison, it now has a population of plains bison and, unfortunately, a large hybrid herd. There are also moose, elk, and caribou. The most vicious creature here is the mosquito. The park is so remote that only a few hundred people visit it each year *(Box 750, Ft. Smith, Northwest Territories, CANADA X0E 0P0)*.

Elk Island National Park, just east of Edmonton, is not truly an island, but a calm area in a sea of bustle. It was formed to protect the last of the prairie elk herd and then deemed the place to save the wood bison, which was once nearly extinct. There is now a population of the bison in a paddock, as well as the elk. There are also moose, mule deer, and whitetails *(Site 4, RR 1, Ft. Sask., Alta., CANADA T8L 2N7)*.

BRITISH COLUMBIA may be the most beautiful province in Canada and certainly has more than its share of wildlife.

Fort Rodd Hill National Historical Park, near Victoria, is not one of the best wildlife parks in the province, but it does have a herd of extremely tame blacktail deer that wander the grounds of the fort *(604 Belmont Rd., Victoria, B.C., CANADA V9C 2W8)*.

Glacier National Park is one of the most rugged areas of western Canada, and the severe weather restricts the numbers of animals that can exist here. Moose and deer are uncommon as are elk. Only mountain goats seem suited to the area and can occasionally be viewed on ridges and cliffs facing the highway. Both black and grizzly bears are here as well *(Box 350, Revelstoke, B.C., CANADA V0E 2S0)*.

Kootenay National Park lies on the west side of the Continental Divide with Banff on the east, and the two parks are similar. Mountain goats can be seen on Mt. Wardle from the highway. Elk and deer range near timberline in the warmer months. Bighorn sheep graze on high meadows, too, and mountain goats graze even higher and come to Radium Hot Springs for the minerals. They spend the winters on grassy slopes at lower elevations *(Box 220, Radium Hot Springs, B.C., CANADA V01 1M0)*.

Yoho National Park is next to Jasper and very similar. Yoho is on the west and Jasper is on the east side of the Continental Divide. The same complete group of mountain monarchs lives here as in Jasper: mule deer, elk, moose, and mountain goats. One of the highest falls in the world, Takakkaw Falls, is also here. Emerald Lake is a particularly scenic area, too *(Box 99, Field, B.C., CANADA V01 1G0)*.

Garibaldi and Golden Ears provincial parks in the wild Coast Range peaks are frequented primarily by fishermen, dedicated backpackers, and mountain climbers. Winter is, they say, three of the four seasons. But here also roam deer and mountain goats *(Parks Branch, 1019 Wharf St., Victoria, B.C., CANADA V8W 2Y9)*.

Mt. Assiniboine Provincial Park lies between Banff and Kootenay national parks, two of the most awesome in Canada. It is difficult to reach—by trail or air—but is home to the great animals for which the area is known everywhere: elk, deer, moose, mountain goats, and bighorn sheep *(1019 Wharf St., Victoria, B.C., CANADA V8W 2Y9)*.

Mt. Robson is the highest peak in the Canadian Rockies, towering over the western park entrance. The park is bordered on the east by Jasper National Park in Alberta. Moose may be seen in the marshes at the east end of Moose Lake, while mountain goats frequent the rock slides on the north side of the highway. Mule deer can be anywhere, and elk roam the eastern section *(Box 579, Valemount, B.C., CANADA V0E 2Z0)*.

Tatlatui Provincial Park encompasses the Atlin-Stikine wildlife management unit. It is difficult to see, most visitors entering by horseback or small plane, but is known the world over for its Stone and Fannin sheep (a cross between Stone and Dall), mountain goats, and moose. Elk are found in some areas, too *(Parks Branch, 1019 Wharf St., Victoria, B.C., CANADA V8W 2Y9)*.

Tweedsmuir Provincial Park is 2½ million acres of rugged untamed wilderness. The alpine meadows and parks, slopes, and open valleys are home to moose, mountain caribou, and mule deer. Elk are making a comeback in the interior but are seldom seen. Access is difficult *(Parks Branch, 1019 Wharf St., Victoria, B.C., CANADA V8W 2Y9)*.

MANITOBA is the site of **Riding Mountain National Park**. Here plant and animal communities of north, south, and west meet, and there is a great variety of plantlife. The park has elk, whitetail deer, and moose. Driving the park roads early and late in the day may help you find them, and there are also ranger-led walks *(Wasagaming, Manitoba, CANADA R0J 2H0)*.

NEW BRUNSWICK is best known to Americans as the site of the **Bay of Fundy**. There is a national park here by the same name where moose and whitetail deer can be seen. The park is on the north shore of the bay where wild seas with a tide of more than 50 feet whip in twice a day sculpting the sandstone cliffs into wonderful shapes *(Bay of Fundy National Park, Alma, N.B., CANADA E0A 1B0)*. Another place to see moose and whitetails is **Kouchibouguac National Park** on the Atlantic coast, which also includes part of the Acadian Forest *(Kent Co., N.B., CANADA)*.

Whitetails live throughout the province. For more details write: *Fish and Wildlife Branch, Centennial Bldg., Fredericton, N.B., CANADA E3B 5C3*.

NEWFOUNDLAND AND LABRADOR, in the most easterly part of Canada, have a national park where both caribou and moose can be found mostly in the interior boreal wilderness, with fewer on the scenic coastal region. Seals can also be seen on the rocky shore, and whales at sea. The park, **Terra Nova**, encompasses some areas accessible only by boat. *(Glovertown, Nfld., CANADA)*.

Newfoundland's provincial parks are scattered throughout the province. Moose and both barren-ground and woodland caribou occur in large numbers in the interior. Many of the best places can only be reached by air, and they are not the easiest places to see the game. Write for *Newfoundland and Labrador Provincial Park Guide (Dept. of Tourism, Parks Division, Box 9340, St. John's, Nfld., CANADA A1C 5T7)*.

NOVA SCOTIA, the province that seems an extension of the U.S. state of Maine, has a large population of whitetail deer, especially in the west where there is dense forest. The best place to look for the deer is in **Kejimkujik National Park** *(Maitland Bridge, N.S., CANADA)*.

ONTARIO, a huge province with endless wilderness, has large numbers of moose and whitetails. Again, the massive areas cannot be covered here, but there are two books available for those who care to brave the far outlying tracts of Ontario. Write for *Ontario Provincial Parks (Information Branch, Ministry of Natural Resources, Parliament Buildings, Toronto, Ont., CANADA M7A 1W3)*, and for another book, *Vacation Planning (Canada Government Office of Tourism, 105 Kent St., Ottawa, Ont., CANADA K1A 0H6)*.

If you're truly adventurous, consider a visit to **Pukaskwa National Park** on the northern shore of 400-mile-long Lake Superior, which is the most rugged park in Ontario. There are moose in this park east of Lake Nipigon in Nipigon Provincial Forest, but access is difficult *(P.O. Box 550, Marathon, Que., CANADA P0T 2E0)*.

Two-million-acre **Algonquin Provincial Park** is well known for its brook trout, but it also has large numbers of whitetails and moose. The park is near Toronto with easy access to some parts. Others are wild and remote and can only be explored by foot or canoe. Park naturalists lead walks during the busy summer season *(Ministry of Natural Resources, Whitney, Ont., CANADA K0J 2M0)*.

Quetico Provincial Park, only half as large as Algonquin, is still spacious and has some of the world's oldest rocks. Forty percent of the park is water, and the land is wild and primitive. Only canoe routes lead to much of it. But there is a good population of moose and whitetails here *(Atikokan District Office, Ministry of Natural Resources, Atikokan, Ont., CANADA P0T 1L0)*.

Lake Superior Provincial Park is wild and generally inaccessible, but has a decent population of moose *(District MNR, Aerodrome Building, Box 130, Sault Ste. Marie, Ont., CANADA P6A 5L5)*.

QUEBEC's reserves cover 51,000 square miles, or three times the size of Switzerland. Much is wilderness with often difficult access *(Reservations Office, Parks Branch, Box 8888, Quebec City, Que., CANADA G1K 7W3)*.

One national park and two provincial parks should be mentioned especially: **Forillon National Park** occupies an eastern section of the scenic Gaspé Peninsula, a thumb of land jutting into the Gulf of St. Lawrence. Look for whitetail deer and moose *(Box 1220, Gaspé, Que., CANADA G0E 1R0)*.

The two provincial parks adjoin, so can be considered as a single unit. These are **Gaspesian** *(Ste. Anne des Monts, Que., CANADA G0E 2G0)* and **Matane** *(263 St. Jerome Ave., Mtne., Que., CANADA G4W 3A7)*. The Schick-Shock (or Chic-Chocs) are the highest ranges east of the Rockies. They are home to moose, whitetail deer, and woodland caribou. In summer look for the caribou in the 4,160-foot-high Mont Jacques Cartier area.

Another provincial park, **Park des Laurentides**, is in the Laurentian Mountains, the oldest in North America. Large moose roam this area, which is divided into two areas: the Kiskissink, which is very wild and in some places can only be seen by canoe; and the Grands-Jardins, a grassy plain of 100 square miles. For a guidebook and more information write: *Ministry of Tourism, Fish,* and Game, 150 E. Boulevard, St. Cyrille, Quebec City, Que., CANADA G1R 4Y3).

SASKATCHEWAN is known for its huge grain harvests, but it also has remnant areas of the original prairies and their native species.

Grasslands National Park was acquired in an agreement between the province and the federal government signed in 1981. It is a large area of mixed native grasses that supports a good population of pronghorns and a number of smaller creatures (Val Marie, Sask., CANADA S0N 2T0).

Prince Albert National Park, in central Saskatchewan, lies between the Rockies and the Canadian Shield; more important, however, is its north-south location. An aspen parkland grows in the south with its smaller prairie species, and there is boreal forest in the north with elk, moose, and caribou, as well as whitetail and mule deer (Box 100, Waskesiu Lake, Sask., CANADA S0J 2Y0).

Cypress Hill Provincial Park lies just 50 miles north of the U.S. border and is a flat-topped plateau. Mostly grassy, there are some woods on the north-facing slopes and in some valleys. It is well-watered. Elk, pronghorn antelope, moose, and whitetail and mule deer can be found here (Maple Creek, Sask., CANADA).

Duck Mountain Provincial Park is known for the beauty of Little Boggy Creek Valley, a mile or two wide and 15 miles long. Deer and moose are here. Mostly in western Manitoba, the address is in Saskatchewan (Box 39, Kamsack, Sask., CANADA, S0A 1S0).

Moose Mountain Provincial Park is an area of aspen with scattered lakes. Whitetail deer live here as well as moose and elk that are otherwise rare in this southeastern corner of the province (Box 100, Carlyle, Sask., CANADA).

YUKON TERRITORY is a huge triangular shape occupying a vast area between British Columbia and Alaska. Its only national park is **Kluane,** easy to reach, located on the Alaska Highway.

Kluane's trails are not maintained but can lead wildlife enthusiasts to the large animals of their dreams. There is a large population of Alaska moose along the flats of the Donjek River, caribou in the Duke River area, and Dall sheep and mountain goats are often seen from the highway on the steep cliffs overlooking the road. Passers-by can stop and see them through a spotting scope at Sheep Mountain on the Alaska Highway. Kluane and Wrangell-St. Elias National Park, its neighbor in Alaska, were proclaimed a United Nations World Heritage Site in 1980 (Haines Jct., Yuk., CANADA Y0B 1L0).

Other areas to see woodland and barren-ground caribou, Dall, Stone, and Fannin sheep, mountain goats, and moose—plus a few deer in the south and west—are the **Fishing Branch Preserve** due north of the Peel River in the far North, the **Peel River Preserve** in the Richardson Mountains, and the **McArthur Game Sanctuary** south of Mayo (Director of Game, Box 2703, Whitehorse, Yuk., CANADA Y1A 2C6).

Oldsquaw Lodge on the remote border of Yukon and the Northwest Territories holds another sort of viewing opportunity. A gathering place for naturalists, moose can be spotted in the willows, and grizzlies roam the open tundra. There is almost never a time when caribou cannot be seen through the lodge's spotting scopes. Smaller creatures live here in summer in vast numbers too, including over 100 species of birds (Sam Miller, Oldsquaw Lodge, Bag Service 2700 (B), Whitehorse, Yuk., CANADA Y1A 4K8).

Since the discussion here centers on the horned and antlered creatures to be seen, we have made little mention of the bears that often share the same habitat. Be warned that they are around and are not to be trifled with. Alertness and care are recommended by the authors.

For a more complete listing of wildlife areas in western Canada, see our book, *Photographing the Wild Northwest.*